THE FLOER JUNGLE

Charting the Development of a Theory

HISTORY OF MATHEMATICS VOL **51**

THE FLOER JUNGLE

Charting the Development of a Theory

Helmut Hofer
Siobhan Roberts

AMERICAN MATHEMATICAL SOCIETY
Providence, Rhode Island

Cover art courtesy of Landon Clay (oil painting, "The Mathematical Jungle").

2020 *Mathematics Subject Classification.* Primary 01A65, 01A70, 37B30, 37J12, 37J39, 53D40, 57R58, 58E05, 58E30.

For additional information and updates on this book, visit
www.ams.org/bookpages/hmath-51

Library of Congress Cataloging-in-Publication Data

Names: Hofer, Helmut, 1956- author | Roberts, Siobhan author
Title: The Floer jungle / Helmut Hofer, Siobhan Roberts.
Description: Providence, Rhode Island : American Mathematical Society, [2026] | Series: History of mathematics, 0899-2428 ; volume 51 | Includes bibliographical references and index.
Identifiers: LCCN 2026008071 | ISBN 9781470485146 paperback | ISBN 9781470485931 ebook
Subjects: LCSH: Symplectic geometry | Floer homology | Mathematicians–Germany | Differential topology | Floer, Andreas, 1956-1991 | AMS: History and biography – History of mathematics and mathematicians – Contemporary | History and biography – History of mathematics and mathematicians – Biographies, obituaries, personalia, bibliographies | Dynamical systems and ergodic theory – Topological dynamics – Index theory, Morse-Conley indices | Differential geometry – Symplectic geometry, contact geometry – Floer homology and cohomology, symplectic aspects | Manifolds and cell complexes – Differential topology – Floer homology | Global analysis, analysis on manifolds – Variational problems in infinite-dimensional spaces – Abstract critical point theory (Morse theory, Ljusternik-Schnirelman theory, etc.) | Global analysis, analysis on manifolds – Variational problems in infinite-dimensional spaces – Variational principles | LCGFT: Biographies
Classification: LCC QA665 .H6425 2026
LC record available at https://lccn.loc.gov/2026008071
History of Mathematics ISSN: 0899-2428 (print); 2472-4785 (online)
DOI: https://doi.org/10.1090/hmath/51

∞ The paper used in this book is acid-free and falls within the guidelines established to ensure permanence and durability.
Visit the AMS home page at `https://www.ams.org/`

10 9 8 7 6 5 4 3 2 1   31 30 29 28 27 26

Dedicated to the Memory of
Andreas Floer, Dietmar Salamon, Kris Wysocki, Eduard Zehnder

Contents

Prologue

Mathematics is an adventure, mathematicians the adventurers—as in the story of Floer theory and its progenitor, the German geometer Andreas Floer. An exceptional mathematician, Floer was complex, aloof, enigmatic. He exhibited, as one friend put it, "radical individuality." He inhabited the world with a singularly critical way of thinking, and a quintessential disregard for convention. This is what allowed him to tackle questions like: In infinite dimensions, do there exist infinite-dimensional holes? His revolutionary ideas, contradicting the established wisdom, can only be understood in this context. There is no doubt that Floer is an important figure in the history of mathematics. The power and reach of his work only continue to grow.

As a graduate student in the early 1980s, Floer was particularly intrigued by symplectic geometry, combining geometry and dynamics. Emerging from classical mechanics, this domain investigates, among other things, mathematical models that describe the evolution of physical systems, such as our planetary system or a satellite's orbit around celestial bodies. When Floer arrived on the scene, symplectic geometry was experiencing tremendous growth, drawing in mathematicians with eclectic intellectual expertise and cultural sensibilities—spurring breakthroughs and attracting more and more enthusiasts. Under the umbrella of symplectic geometry there now exists symplectic topology and symplectic dynamics, both of which have contributed to the solution of longstanding problems and motivated new questions. The abundance provided new tools for other fields of inquiry, such as low-dimensional topology, dynamical systems, and algebraic topology, as well as physics.

Floer possessed extreme self-confidence in mathematics, and in the midst of this lush landscape he flourished. But Floer lacked more practical, everyday self-assuredness. He suffered from depression. He failed to find firm footing in life. In 1991, at the age of 34, he committed suicide. The tension between his personal and mathematical lives set the stage for his success, and his downfall. His friend Rolf Kaiser, in trying to explain Floer's "catastrophe," borrowed from Oscar Wilde: "The real fool, such as the gods mock or mar, is he who does not know himself."

The Floer Jungle celebrates Andreas' journey and the path of those who traveled with him—before, during, and after. It aims to give a glimpse of what doing mathematics looks like—the messy, thrilling, often overwhelming process—and how the people who do it live, interact, rejoice, despair.[1]

Lay of the Land

One approach to mathematics—rarely if ever taken by Floer—is to imagine that you're standing in front of a big wall: The problem. You intend to break through the wall by any means necessary. Maybe a sledgehammer will suffice; perhaps heavy artillery is needed. But really, it's just you against the wall. You run hard at it, again and again. The wall does not budge, but if you're powerful enough, you create a crack. You run at it some more. The crack gets bigger. Eventually you might get through. That's a success story. Your head hurts, but the adrenaline overrides the pain.

Another strategy—Floer's preferred path—is to imagine venturing into a jungle, depicted in Landon Clay's oil painting, see page xiv. This is a realm populated by mathematical flora and fauna, the subjects under investigation. Out on daily field work, you hack away at encroaching brush and dangling vines. You're clearing the way, making progress; sometimes halting, amid many hazards. Yet slowly but surely there is forward momentum. Sometimes, turning a corner or mounting a peak, valleys and vistas suddenly reveal themselves, alive with unexpected points of interest.

Over the years, many have heard Helmut Hofer's jungle metaphor. To wit, in November 2016, a formative discussion about this book took place over lunch at the Institute for Advanced Study in Princeton. Joining at the lunch table that day was the mathematician Nathaniel Bottman—he became a group leader at the Max Planck Institute for Mathematics in Bonn; and from August 2026 he is an assistant professor at the University of Vermont. As the founding CEO and principal scientist at a startup called Incubilate, Bottman directs his expertise—acquired while developing schemes for data organization in Floer theory—at "An Operad-Based Framework for Deep Learning," which is part of the DARPA Artificial Intelligence Quantified program. For Bottman, the jungle metaphor resonated—he'd heard this narrative of Floer's approach before, and he extrapolated on the idea in a follow-up email. "Sometimes it's a slog and you spend a lot of time exploring a boring tributary or get mired in mud," he said. "But other times you make a discovery that you never expected—maybe you cross a ditch, and an hour later you cross another ditch, and after crossing a bunch of ditches you realize that all these ditches are part of a huge pattern that some ancient civilization dug into the earth, like the Nazca Lines"—geoglyphs in Peruvian desert sands. Furthermore, Bottman offered a reinterpretation of the jungle philosophy as applied to his own work: "When I come across a new object, I try not to impose any rules or properties that I think it should have (oh, there are a bunch of creek-beds here because we're in the lowlands and near a river)," he said. "Rather, I imagine that this object has some fundamental form, and I should try to understand the object well enough that I can uncover its purest realization (Nazca Lines). And this point of view has helped me several times."

Unlike the problem-oriented approach, in the jungle you're not fixated on any one creature. Instead of running against that big wall, you wander right, meander left. Then finally, out of nowhere, you stumble upon—of all things—a door. It's unlocked and opens onto the same place: the other side of the wall. Not only have you achieved what you wanted and arrived at your desired destination, you've also learned about lots of unexpected entities along the way.

* * *

This is not a textbook on Floer theory; neither is it meant to be historically exhaustive. Rather, the goal is to retrace Floer's path—as it winds through the landscape—and recreate something of the "Floer experience." We chart the development of ideas that led to Floer's seminal contributions; and at once the sociological evolution of what ultimately became known as Floer theory. The story offers an insider's view, chronicling a compelling development in modern mathematics. In a essence: Floer invented a scheme for counting so-called critical points of functions on infinite-dimensional spaces.

FIGURE 1. The Mathematical Jungle, Landon Clay (2024).

The narrative is directed at mathematically grounded readers—graduate students in mathematics, physics and engineering; researchers of various stripes—and more generally those who are seriously scientifically curious. We reconstruct the mathematical terrain, the existing ideas, and introduce the main characters, describing their interactions, their biases and judgments, wrong or right. We provide supplementary appendices, footnotes and references so that the reader can explore and build up knowledge. Basic prerequisites are multivariable calculus and linear algebra; exposure to ordinary differential equations and differential geometry of surfaces will prove useful too—but worry about it when you get there. There will be a junction at which more junior readers will realize that they lack the knowledge to proceed. But this is not uncommon in research—as Peter Sarnak, of the Institute for Advanced Study and Princeton University, has been known to say: "The steady state of a mathematician is frustration." So, be comfortable with the discomfort, with the fact that you need to learn as you thrash through the jungle.

The Floer Jungle proceeds with an intense mathematical tale. Most of the exposition is by example and illustration rather than formalism. It is perfectly fine to skip some of the more challenging parts, or to pause and read up on the missing concepts—every research mathematician has to do so constantly—and then continue.

FIGURE 2. Andreas at about age 20, 1976.

Interwoven throughout the narrative are various dimensions of Floer's story, memories collected from family, friends, and colleagues that sketch out a portrait of the inimitable man who made an unexpected contribution. In this sense, the book is perhaps an unlikely hybrid: at once an expository workbook that entices the reader to engage with intriguing mathematics, and a biographical memoir that invites a more intimate consideration of mathematicians' lives.

* * *

Within the realm of symplectic geometry—which generally is considered the geometry of the principle of the least action[2]—Floer's ideas transformed the centuries' long advancement of critical point theory. For instance, minima and maxima are two types of critical points.

Influenced by Isaac Newton's laws of motion, Pierre Louis Maupertuis introduced the "principle of least action." He observed that in a physical system time evolution happens in such a way that a certain mathematical quantity—the "action"—is minimized, or maximized. Joseph-Louis Lagrange solidified these insights with his treatise about analytical mechanics. Leonard Euler evoked the least action principle in laymen's terms as follows: "For since the fabric of the universe is most perfect and the work of a most wise Creator, nothing at all takes place in the universe in which some rule of maximum or minimum does not appear."

As the natural sciences progressed, it became clear that Maupertuis' principle of the least action could be generalized to the principle of "stationary action"—where, besides minima and maxima, other critical points emerged.[3] This generalization, however, came with a price: Practically, how could such critical points be found? While methods for finding minima and maxima of certain action functionals had been developed—and allowed the prediction of useful orbits for classes of dynamical systems—some of the critical point problems could not be studied. Mathematicians wondered whether there was key mathematical machinery missing, yet to be developed. Worse, they wondered whether the principle of stationary action was simply useless from a practical perspective. They assumed the latter, and they were wrong.

Floer found a door that opened onto a beautiful landscape. Suddenly, the useless and the hopeless became tractable. He devised the techniques, the machinery, to find entities of interest. And he found robust ways to count critical points in infinite dimensions—a counting method that essentially deformed a difficult problem into an easier problem.

Through careful analysis, Floer realized the big picture. His story is not about a flash of genius. It is broader and deeper than that. It's about wild ingenuity transforming a complicated and messy landscape into a powerful theory: Floer theory. This breakthrough inspired further advancements that continue to this day. Rohil Prasad, a former Ph.D. student of Hofer's at Princeton, then a Miller Research Fellow at Berkeley and currently an assistant professor at Princeton, can hardly fathom mathematics without it. "Whenever I'm considering a new problem, one of the first thoughts I have is, in some form, 'What can Floer theory say about this?'." He continued, "The 'quantitative' impact of Floer theory is clearly described by the theorems which it has been able to prove, but a description of its 'qualitative' impact, how it has changed the way we see and describe symplectic spaces, is more difficult for me to express. Floer theory forms the bedrock of my mathematical training, so I am still learning how people thought before its introduction."

Floer's Deadly Sin

To set the scene at a pivotal moment in time: In October 1986, some thirty mathematicians convened for a NATO Advanced Research Workshop in Il Ciocco, Italy—from the lecture room, participants stepped onto a terrazza and into a vista of rolling Tuscan mountains. Although the invitation list was select, it was a big event. Attendees included researchers in dynamical systems and celestial mechanics and nonlinear analysis, geometers and topologists and mathematical physicists—all were excited and intrigued by the prospects emerging in the realm of symplectic geometry.

Andreas Floer was invited but couldn't attend. He was just settling into a new job at the prestigious Courant Institute for Mathematical Sciences at New York University. That summer, however, Floer had explained the details of a proof—his proof that would blossom into Floer theory—to his former Ph.D. advisor, Eduard Zehnder, an analyst and dynamicist. Zehnder, in turn, brought word of this startling new research to Il Ciocco.

One morning Zehnder went for a walk with the formidable geometer Mikhael Gromov, then at Institut des Hautes Études Scientifiques in Bures-sur-Yvette. They talked about Floer's work. When Zehnder arrived at the lecture hall, Paul Rabinovitz, the gathering's organizer, suggested Zehnder give a talk on Floer's new theory. "I said, 'I have no manuscript, but I can tell you what I know,' " Zehnder recalled.

Zehnder delivered a blackboard talk in his typical style, but the response from the audience proved atypical.

He began with a lot of definitions: Morse theory, Lagrangian transversal intersections. "I introduced all the concepts, which took a while. And then I came to the main part, namely the Morse equation of these intersections. Then suddenly I heard a scream in the audience. Somebody jumped up, rushed to the blackboard, took away my chalk."

It was Gromov. He was not happy about what he'd been hearing. He pronounced: "I have a counterexample!" He turned to the blackboard and wrote it down. "When he stepped aside, I saw that it was just little circles, very wild. I didn't understand what the counterexample was. I asked the audience whether anybody understood the counterexample. There was silence." Gromov was a great mathematician. Floer was clearly wrong. Zehnder finished his talk. "I really felt like a fool," he said.

In short, Gromov deemed Floer's ideas to be nonsense. Specifically, regarding the relevant matter of pseudoholomorphic curves, they possess symmetries and in Gromov's view symmetries always exist for good reason. Floer had destroyed the symmetries—an action from which, Gromov believed, no good could come.

Some ten years later, when Gromov received the 1997 Steele Prize, he rehashed the point about Floer's symmetry breaking. "I still cannot forgive him for this," he said. He added: "Alas, prejudice does not pay in science." In a 2022 "Zoominar," [**H2**], Gromov's initial skepticism of Floer's work was revisited. Hofer speculated about the mathematical underpinnings of his initial rejection. In response, Gromov reiterated that is was the symmetry breaking and with a laugh he added: "I still don't like it."[4]

Because, of course, it turned out that Gromov was wrong. Floer's symmetry breaking proved powerful.

* * *

To bring into view the broader context: Generally, variational principles refer to a situation where the solution of an equation can be understood as a critical point of a function defined on a finite or infinite-dimensional space. Many of these principles play important roles in mathematics, physics and biology—sometimes even unearthing philosophical questions.[5] However, the Floer story centers around

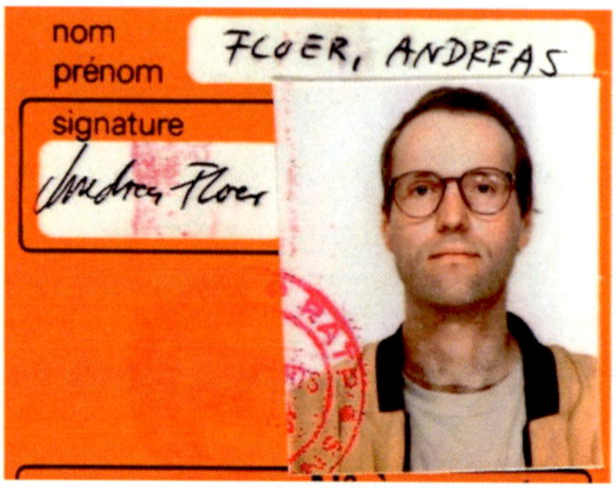

FIGURE 3. Andreas, circa 1990.

a class of variational principles that held some formal significance in physics but was considered useless for mathematical purposes.

To explain the ideas, let's consider a physical system whose set of possible states defines an infinite-dimensional space which we shall call Λ. A function is defined on Λ which associates to a state a number, interpreted as its energy, and written as

$$E : \Lambda \to \mathbb{R} : S \to E(S).$$

Such an infinite-dimensional space often allows a version of calculus (with infinitely many variables). In this case we are interested in the states S in Λ where the derivative $dE(S)$ vanishes:

$$dE(S) = 0. \tag{1}$$

Such points S are called critical points.[6] In the relevant cases within mathematics and physics, the equation (1) is very often a differential equation describing a physical law.

Having a variational principle for certain types of equations means that there exists an energy functional E such that the solutions of the equation can be identified with critical points of E. In practice there is a huge difference between solving a general equation $F(S) = 0$ compared to one of the special form $dE(S) = 0$. The mathematical toolbox for solving the latter is much larger; and finding critical points is often an easier task due to the availability of what mathematicians call variational methods. Floer theory—in some sense—is a significant extension of the variational viewpoint and toolbox.

In physics, a ground state is a state with the smallest energy. As in one-variable calculus, where the derivative vanishes at a minimum, ground states are critical points; and the vanishing of the derivative often means that they satisfy a differential equation. These ground states are in line with Euler's remark that "... nothing at all takes place in the universe in which some rule of maximum or minimum does not appear."

One of the earliest known variational principles is Pierre de Fermat's principle of least time, which he proposed in 1662. In a non-homogeneous glass-like medium the speed of light varies depending on the local change of the media properties. Having two points A and B lying in the medium, Fermat observed that a beam of light traveling from A to B takes the path that minimizes the time of travel. As Richard Feynman explained it: Think of a life guard trying to save a drowning swimmer. The life guard has to deal with two possible media types, namely sand

and water. Assume he can run three times as fast on sand than swim in water. With A being the position of the life guard's post and B the position of the swimmer, depending on the locations of A and B the life guard will first run along the beach before jumping into the water in order to get to the drowning swimmer as fast as possible.

However, in many situations variational principles are considered useless for practical purposes. They seem formal and cannot be exploited by the application of variational method since certain necessary conditions for applying the machinery are not met. This was true for many of the variational principles arising once Maupertuis' least action principle had been extended to the principle of stationary action.

In the Floer story, the variational principles under consideration had not only been deemed useless, but their uselessness seemed to be underpinned by an *almost* watertight argument. Floer turned these seemingly useless structures into powerful tools.

The transformation did not arrive out of the blue. There is an impressive cast of characters who broke various trails: The foundation was laid by Marston Morse, followed by Richard Palais and Stephen Smale, and continued with Jürgen Moser and Paul Rabinowitz. Charles Conley and Eduard Zehnder developed ideas further and achieved a breakthrough on a longstanding problem. Other threads include Simon Donaldson, Michael Freedman, Clifford Taubes and Karen Uhlenbeck around gauge theoretic methods in four-dimensional geometry; Edward Witten's approach to Morse Theory; Misha Gromov's seminal contributions to symplectic geometry, as well as Yasha Eliashberg, who got the ball rolling, with John Mather and Anatole Katok in supporting roles. And an indirect contribution, which undoubtedly shaped part of the existing landscape, is Ivar Ekeland and Jean-Michel Lasry's result on the number of periodic orbits on a convex energy surface, [**EL**], and Ekeland's Morse theory for such energy surfaces, [**Ek1**].

From this messy landscape Floer set forth, gradually clearing a vista.

The eminence grise Sir Michael Atiyah, reflecting upon Floer and his ultimate feat, once made a comparison with Alexander Grothendieck: "I remember Grothendieck was a whirlwind, he talked for hours nonstop—that was an experience never to be forgotten," Atiyah said. "Floer was the other end of the spectrum—there were more reasons to forget. He was a quiet person, subdued, he didn't make a big impact." As for Floer's theory, that was another matter. Atiyah, ever the mathematical matchmaker, was keen to spread the theory's impact around. "It was Fields Medal material, definitely; it was up to the Fields standard," said Atiyah. Atiyah admitted to doing a lot of propaganda at conferences in those days; he viewed his main role as trying to explain new developments. "People thought I kept bad company because I mixed with physicists. Physicists were too woolly," he said. "But Floer had spectacular new ideas that they could use, and at just the right moment. He bridged a gap—he provided stepping stones—between mathematics and physics. Mathematics comes in fits and starts and here was a brand new idea that came from nowhere, or seemed to, and opened a door to the future, and a period of heightened activity."

"That's rare," said Atiyah—but inspired by others, Floer opened such a door, and then "went out and explored all the wonderful new trees and flowers."

Welcome to the Floer jungle.

Part 1

Foundation

CHAPTER 1

Birth-Death

Raoul Bott[7] once delivered a lecture titled "Morse Theory Indomitable."[8] Bott recalled that when he first met the great Marston Morse [**Bo2**], "he preached the gospel of 'critical point theory' "—a.k.a. Morse Theory—"first, last and forever, to such an extent, that we youngsters would wink at each other whenever he got started." As a professor at the Institute for Advanced Study from 1935 to 1962, Morse was also renowned for the parties he and his wife Louise held for new members at the beginning of each semester.

FIGURE 1. Marston Morse.

"But it is time to get down to business," Bott continued, "and what I would like to do here today is to try and sort out the various stages of Morse Theory I have lived through ...."

1.1. Pictures in Dimension One

Critical point theory and its powerful methods have proved important not only for different fields of mathematics, but also for physics and, more recently, data science. We shall illustrate the ideas in some easier cases and try to convey the underlying spirit. Starting with one-dimensional cases, we take a deeper look at aspects of one-variable calculus.

1.1.1. One-dimensional Landscapes. Consider one-dimensional landscapes as shown in Figure 2. Each landscape can be viewed as the graph of a function[9] f which associates points on the green segment (the sea level) with corresponding heights on the landscape above.

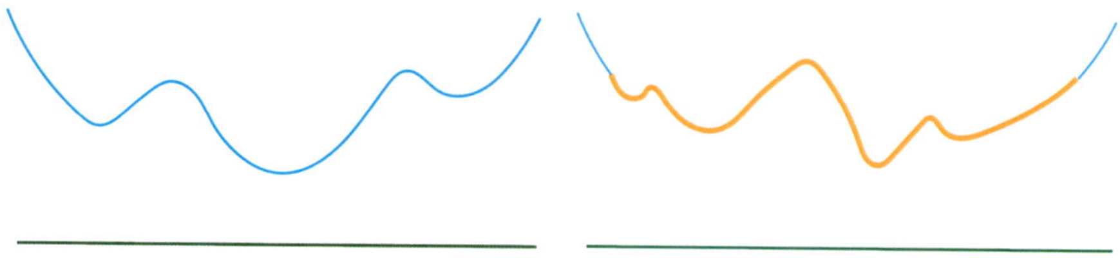

FIGURE 2. Each of these landscapes are the same at the outer edges, but the interiors differ.

We assume that the graphs are generic in the sense that the local valleys and peaks are well-formed.[10] If we travel on our landscape we must climb over peaks and walk through valleys. In the landscape pictured at left we have two peaks and three valleys and we form the number

$$\#(valleys) - \#(peaks) = 3 - 2,$$

which takes the value 1. Computing the same expression as before for the landscape at right we obtain

$$\#(valleys) - \#(peaks) = 4 - 3 = 1.$$

We obtain the same number as before despite the fact that we modified the interior landscape. This is not an unexplainable coincidence since in the background of this regularity lies what is called the birth-death mechanism. This allows us to count critical points in a fancy algebraic way. The birth-death mechanism, as exhibited in our one-dimensional examples, is the shadow of a deep fact which holds true in all finite dimensions.

From nothing, we can create two critical points, namely a valley and a peak, and similarly we can have two such critical points collide and cancel each other out, see Figure 3.

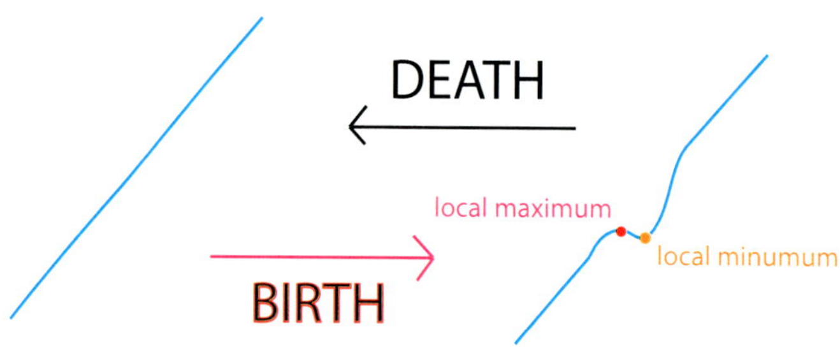

FIGURE 3. The birth-death mechanism, shown schematically.

Let us consider another landscape and observe that the behavior of our new function f near the boundary of the green interval is different from the previous cases. In Figure 4 we compute

$$\#(valleys) - \#(peaks) = 5 - 5 = 0$$

and modifying again, see Figure 5, by keeping the landscape fixed at the outer edges we compute

$$\#(valleys) - \#(peaks) = 0 - 0 = 0.$$

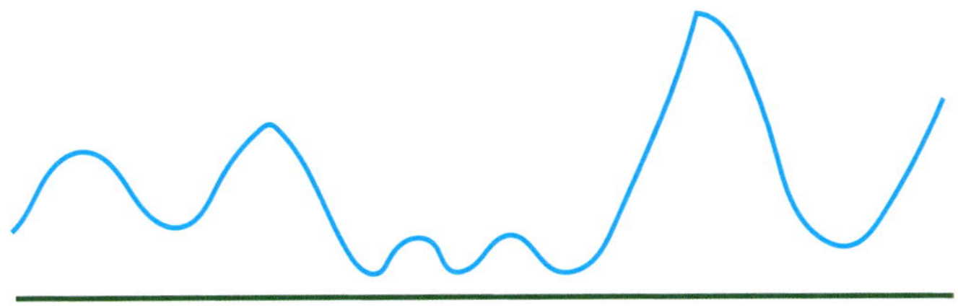

FIGURE 4. The landscape is different at the outer edges.

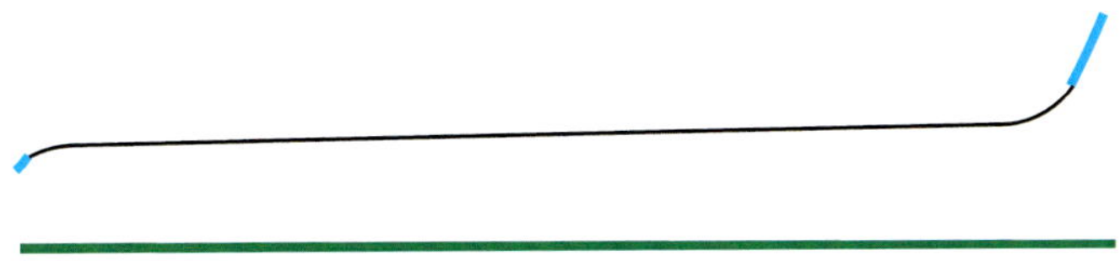

FIGURE 5. In this case all the critical points could be eliminated.

The upshot is that the number we obtain, an algebraic count, only depends on the behavior at the outer edges of the landscape. In the background is a relationship between topological phenomena and calculus. This way of producing some numbers is "algebraic topology light." Our counting rule has been that a valley counts 1 and a peak counts -1. If we are more careful we can obtain an even more detailed description of the situations we have discussed before. This description will exhibit, in an algebraic way, the birth-death mechanism as well.

1.1.2. Packaging Data and Counting. Given a function f with well-formed valleys and peaks we assume that the behavior at the far edges is as in our previous examples, namely there are no valleys and no peaks. We would like to encode data about f with the goal to turn our previous discussion into a mathematical statement.

There are in general different ways to describe a mathematical object. Depending on the context, one description might be better than the other, even when the descriptions in some sense are equivalent. Also one might use a mathematical description which just carries enough information for the question one likes to study. While one can argue that mathematical structures exist in the universe independent of mathematicians, it still rests on the skill of mathematicians to find (or construct) these structures and describe them. For a variety of viewpoints on this philosophical question see the discussion by an engineer and physicist [**Ab**] in a paper entitled "The Reasonable Ineffectiveness of Mathematics."

We shall pick a particular algebraic description because it is seemingly convenient. Fortunately it also turns out to be a good one when we emphasize the importance of the birth-death mechanism.

The reader should recall the notion of a polynomial in one variable t, which is a formal expression of the form

$$P(t) = a_0 + a_1 \cdot t + \ldots + a_n \cdot t^n,$$

where $a_0, \ldots, a_n$ are real numbers and t is a variable. Two polynomials can be added, subtracted and multiplied. There is even an algebraic long division of a polynomial by another polynomial—already known to the Greeks—performed by the Euclidean algorithm.[11] We can also evaluate a polynomial $P(t)$ at any number t.

Denote by a_0 the number of strict local minima and by a_1 the number of strict local maxima.[12] We associate to f the simple polynomial (of degree at most 1), called the Poincaré polynomial associated to f,

$$P_f(t) = a_0 + a_1 \cdot t.$$

That we encode the critical point data associated to f in the form of a polynomial has at this point no other reason than convenience—the variable t after a_1 just reminds us that a_1 is a count of local maxima and similarly $a_0 \equiv a_0 \cdot 1 \equiv a_0 \cdot t^0$ tells us that we count local minima. In Figure 2 we obtain for the left and right landscapes

$$P_f(t) = 3 + 2t \qquad P_f(t) = 4 + 3t. \tag{2}$$

We already note here that if we evaluate these polynomials at $t = -1$ we obtain in both cases the value 1, which is the $(\#(valleys) - \#(peaks))$ count. If we evaluate at $t = 1$ we obtain in the first case 5 and in the second case 7. This is in fact the total number of critical points, i.e the sum $\#(valleys) + \#(peaks)$.

That our choice of mathematical description is a good one becomes apparent through the following observations. In both cases in (2), we can write

$$P_f(t) = 1 + (1 + t) \cdot Q_f(t), \tag{3}$$

where $Q_f(t)$ is a constant polynomial, i.e. $Q_f(t) = q$ with q being a non-negative integer:

$$\begin{aligned} 3 + 2t &= 1 + (1 + t) \cdot 2 \quad \text{in the left case} \\ 4 + 3t &= 1 + (1 + t) \cdot 3 \quad \text{in the right case.} \end{aligned} \tag{4}$$

This has the following interpretation: The 1 on the right side reflects the fact that all our admissible f have a particular behavior far out right and left; and, moreover, the fact that with the prescribed asymptotic behavior of f there always has to be at least one local minimum. However, if we wiggle a lot we create critical points in pairs, namely a local maximum and a local minimum, which is accounted for by $(1 + t) \cdot Q(t)$. The constant polynomial $Q(t)$ is just the number of such pairs. The equality (3) is a particular case of what are called Morse equalities, which are a refinement of the famous Morse inequalities.

Applying the previous procedure to Figure 4 and Figure 5 we obtain in the first case

$$5 + 5t = 0 + (1 + t) \cdot 5$$

and the second case

$$0 = 0 + (1 + t) \cdot 0.$$

This reflects the fact that a function with the prescribed behavior does not need to have critical points, but when it does have critical points they appear in a number of pairs.[13]

1.1.3. A More General Space.

Next we introduce a further generalization where the function f is defined on a more general space. Mathematicians have introduced a class of spaces, smooth manifolds, which come with a version of calculus defined on them. There are many such spaces in higher dimensions and even infinite-dimensional examples (calculus with an infinite number of variables). For our purposes, it is enough that the reader is aware that smooth manifolds are important not only in mathematics but also in applications such as physics. The availability of calculus on these spaces makes it possible to consider differential equations, tools to define systems which obey a physical law.

In dimension one the essential new example, besides the real line $\mathbb{R}$ and intervals, is the circle $\mathbb{S}^1$.

Figure 6 shows a landscape over the (blue) circle. We view the circle as sea level and the landscape is shown in red (the landscape has regions above and below sea level). Mathematically the landscape can be viewed as the graph of a map defined on the circle.

Mathematicians often denote the circle by $\mathbb{S}^1$ and having a map on the circle which takes real values is noted as $f : \mathbb{S}^1 \to \mathbb{R}$.

Again it makes sense to talk about having well-formed valleys and peaks. For such landscapes, when well formed, we again have a relationship of the form

$$P_f(t) = (1+t) + (1+t) \cdot Q_f(t) =: P_{\mathbb{S}^1}(t) + (1+t)Q_f(t).$$

The term $P_{\mathbb{S}^1}(t) = 1 + t$ just tells us that a landscape over the circle $\mathbb{S}^1$ can be found with precisely one highest and one lowest point, the minimum and maximum. If we wiggle, we gain in addition pairs of critical points through the birth-death mechanism. In our case described in Figure 6 the Poincaré polynomial for our function is $P_f(t) = 4 + 4t$ and we know that $P_{\mathbb{S}^1}(t) = 1 + t$. The Morse equality accounts for all of this as

$$4 + 4t = (1+t) + 3(1+t).$$

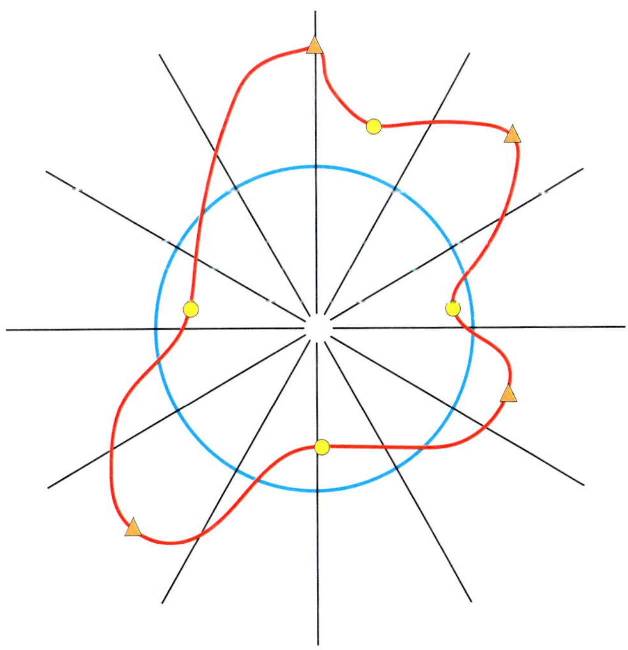

FIGURE 6. A landscape over the circle $\mathbb{S}^1$ with eight critical points—four maxima (orange triangles) and four minima (yellow circles).

1.2. Pictures in Dimension Two

The passage from dimension one to dimension two is a formidable step, yet it can be visualized.

1.2.1. Landscapes. In dimension two, considering the landscape associated to a function, we still see valleys and peaks. In addition we see a new type of phenomena, namely mountain passes.

Here we assume that the behavior of the function f near the critical points is sufficiently nice, i.e. is well-formed. Then, assuming that f is defined in a neighborhood of $0 \in \mathbb{R}^2$ and $f(0) = 0$ and $Df(0) = 0$, one can show—it is called the Morse lemma—that up to a change of coordinates there are three different qualitative behaviors, a valley, a saddle and a peak:

$$f(x,y) = x^2 + y^2 \quad \& \quad f(x,y) = x^2 - y^2 \quad \& \quad f(x,y) = -x^2 - y^2.$$

We can characterize the behavior of a critical point by what is called the Morse index. It is the maximal dimension of a linear subspace $V \subset \mathbb{R}^2$ so that $f|V$ is negative definite.[14] We shall denote a Morse index by m and sometimes indicate the associated critical point. In the three cases at hand we have from left to right: $m = 0, 1, 2$.

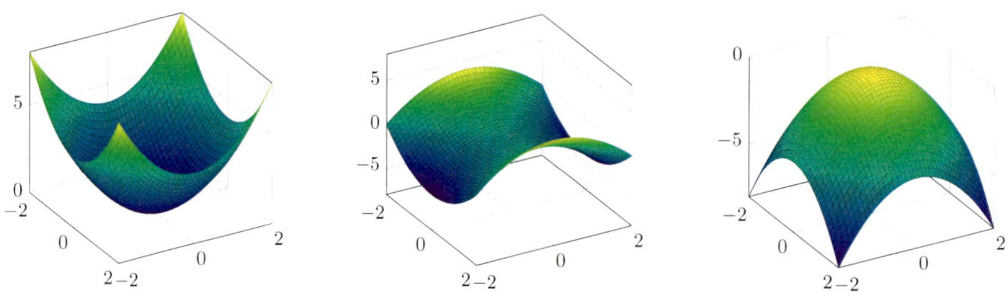

FIGURE 7. From left to right, a local minimum, a saddle point, and a local maximum.

For the first step, as in the one-dimensional case we shall creatively count critical points in more global situations. The rule in dimension two is that we count peaks and valleys as 1 and passes as -1. Recall that in dimension one we counted local minima as 1 and peaks as -1. This looks like a different counting convention, but this is not the case. In any finite dimension the counting rule given above is $(-1)^m$, where m is the previously introduced Morse index. This particular choice of counting again reflects that the birth-death mechanism, which is a generalization of the one-dimensional case, must be accounted for.

The possible cancellation pairs of two critical points are a local minimum and a saddle point or a saddle point and a local maximum, see Figure 8.

A surface which we might think of as a landscape over part of the plane where the Morse function is the height function is shown in Figure 9. In this case we can think of the gradient lines as the paths water drops would take under the gravitational force. This viewpoint is useful when we consider Morse functions on two-dimensional surfaces.

Consider rigid flow lines between two critical points, say x and y. For our purposes "rigidity" means that if we take any point on the line it will flow in

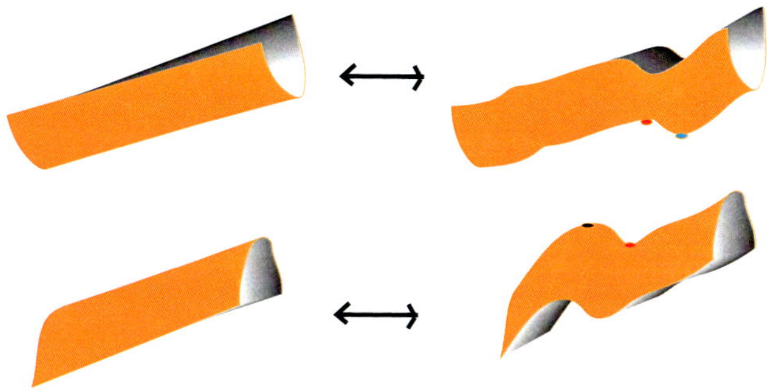

FIGURE 8. The smooth function is the height function, where we assume that the surfaces are embedded in $\mathbb{R}^3$. Shown at top, a degeneration into a saddle point and a local minimum; and at bottom, into a local maximum and a saddle point. The critical points are indicated by colored dots.

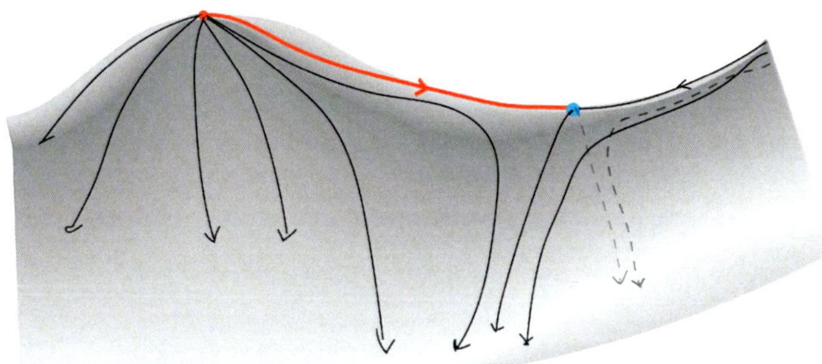

FIGURE 9. A two-dimensional landscape showing a local maximum, a saddle point and some flow lines. From the picture we can identify one rigid flow line (bold red). Others might also be rigid, but we need to know where they end and come from.

forward and backward time to x or y, whereas a point not lying on it will flow in forward or backward time to a critical point different from x and y.

We might project the flow picture onto the plane and obtain Figure 10, referred to as a water flow picture.

Next we consider three water flow pictures on the disk. They are all the same near the boundary, but are different on the insides. In the nice valley on the left of Figure 11 the water flows somewhat radially towards the center. The middle landscape is modified but fixed near the boundary—now we have two valleys, and a pass acting as a watershed in between. We have indicated the lowest points in the valley and the highest point of the pass.[15]

The landscape on the right in Figure 11 has one peak, three valleys and three passes. In all cases we compute the value of $\#(valleys) - \#(passes) + \#(peaks)$ (from left to right)

$$1 - 0 + 0, \quad 2 - 1 + 0, \quad 3 - 3 + 1$$

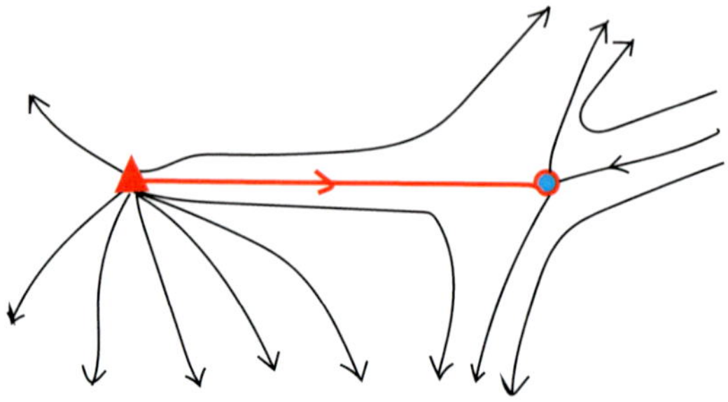

FIGURE 10. The projected flow lines from Figure 9 (including those on the far side the landscape).

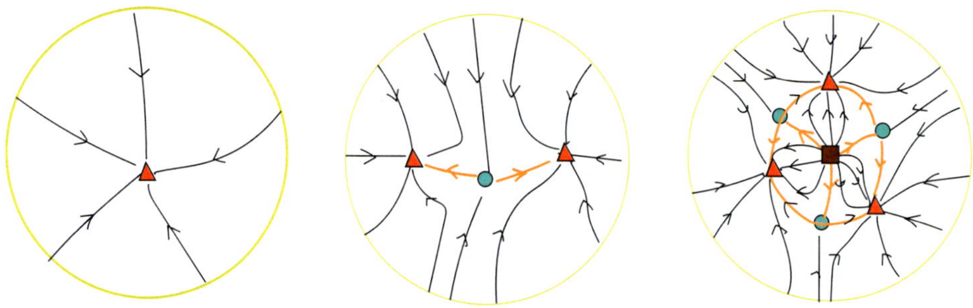

FIGURE 11. Three water flow configurations over the disk. The bold orange arrows indicate rigid flow lines. Red triangles are local minima, blue dots are saddle points and the brown square is a local maximum.

which always gives the value 1. Again there is topology and the birth-death mechanism in the background. Using the polynomial description, where we denote by a_0, a_1 and a_2 the number of valleys, passes and peaks, we define

$$P_f(t) = a_0 + a_1 \cdot t + a_2 \cdot t^2.$$

From left to right we obtain for P_f

$$P_f(t) = 1 \quad \& \quad P_f(t) = 2 + t \quad \& \quad P_f(t) = 3 + 3t + t^2.$$

We find with the situations described in Figure 11 the following relationships going from left to right:

$$\begin{aligned} 1 &= 1 + (1+t) \cdot 0 \\ 2 + t &= 1 + (1+t) \cdot 1 \\ 3 + 3t + t^2 &= 1 + (1+t) \cdot (2+t). \end{aligned}$$

The Q-polynomials are $Q_f(t) = 0$, $Q_f(t) = 1$ and $Q_f(t) = 2 + t$ which all have non-negative coefficients. The 1 in the above formulae represents the minimal number of critical points a function with the prescribed (asymptotic) behavior must have. An example with the smallest number of critical points is given in Figure 11 on the left. The expression with the $Q_f(t)$ describes what can come out from "wiggling."

The identities we have written down are the Morse equalities. The structure is

$$P_f(t) = P_{\min}(t) + (1+t)\cdot Q_f(t), \tag{5}$$

where the polynomial $Q_f(t)$ has non-negative coefficients, which means—writing $Q_f(t) = b_0 + b_1 \cdot t + b_2 t^2 + \ldots$—that the occurring b_i are integers satisfying $b_i \geq 0$. Further P_f is the polynomial count of critical points. $P_{\min}(f)$ is a term describing, in some sense, the minimal number of critical points given the asymptotic behavior of f and $(1+t)\cdot Q_f(t)$ are the contributions from extra wiggling.[16]

For example, in the case shown on the right of Figure 11 the identity is given by

$$3 + 3t + t^2 = 1 + (1+t)\cdot(2+t).$$

It gives the following information, where we expand the right-hand side

$$1 + (1+t)\cdot(2+t) = 1 + 2\cdot(1+t) + t(1+t) = 1 + 2\cdot(1+t) + 1\cdot(t+t^2).$$

The P_f count of critical points in our example comes from a minimum which always has to be there (1); two pairs of critical points, where one pair consists of a minimum and a pass represented by $(1+t)$; and one pair consisting of a (different) pass and a maximum $(t+t^2)$. The latter reflects, of course, the fact that in dimension two we might have different birth-death cancellation pairs, see Figure 8.

1.2.2. More General Examples. When we consider a smooth function on a smooth manifold we can say that each critical point creates or cancels topology. Here topology refers to the fact that spaces can have different properties. This becomes apparent if, for example, we consider the surface of a soccer ball or the surface of a doughnut—they look quite different and mathematicians have developed quantitative tools to measure differences. As a consequence we must have at least as many critical points as the topology of the underlying space demands. Morse theory makes this vague statement quantitatively rigorous with the tools provided by algebraic topology using analysis. The procedure transfers information about the shape of the space to an algebraic description by analyzing any given reasonably nice function on this space.

The first example, see Figure 12, shows two spherical surfaces where we use the previously developed two-dimensional ideas. Both pictures can be viewed as landscapes over a round sphere in analogy to the case depicted in Figure 6.

Then the rule, which reminds of Euler's formula, will be

$$\#(valleys) - \#(passes) + \#(peaks) = 2.$$

But one can say even more: $\#(valleys) \geq 1$ and $\#(peaks) \geq 1$. In fact, one can show that we have, introducing

$$P_f(t) = \#(valleys) + \#(passes)\cdot t + \#(peaks)\cdot t^2,$$

as before an identity of the form

$$P_f(t) = 1 + t^2 + (1+t)\cdot Q_f(t),$$

where $Q_f(t) = q_0 + q_1 t$ and $q_i \geq 0$.

In Figure 13, for the sphere on the left we have the identity

$$1 + t^2 = 1 + t^2 + (1+t)\cdot 0$$

and on the right

$$1 + t + 2\cdot t^2 = 1 + t^2 + (1+t)\cdot t.$$

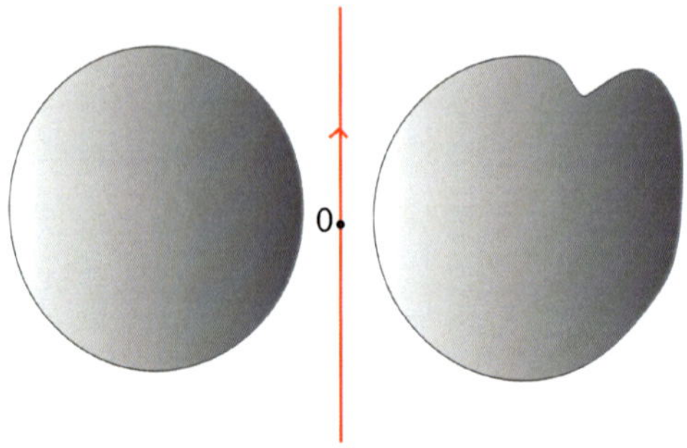

FIGURE 12. On the left we have the round sphere of dimension two (lying in the three-dimensional space $\mathbb{R}^3$); and on the right an indented sphere. We consider the height function h on both of them. With the round sphere, h has precisely two critical points, the global minimum and the global maximum. The indented sphere has four critical points, a global minimum, a global maximum, a local maximum and a saddle point.

FIGURE 13. Here we see two flow pictures. The left picture shows a sphere with precisely two critical points—a minimum and a maximum. The flow lines connecting them lie in a one-dimensional family. On the right we have four critical points. The bold black flow lines are rigid. The thin blue flow lines lie in one-dimensional families which split at the ends into a bold black broken flow line consisting of a consecutive pair of rigid flow lines.

For a landscape over the surface of a doughnut we have the formula

$$\#(valleys) - \#(passes) + \#(peaks) = 0.$$

To show such relationships in full generality is already difficult and requires more sophisticated mathematical machinery.

In the torus case, see Figure 14, the Morse equality turns out to be of the form

$$P_f(t) = 1 + 2t + t^2 + (1+t) \cdot Q_f(t). \tag{6}$$

If we evaluate the above equality at $t = -1$ we see that the term $(1+t) \cdot Q_f(t)$ is killed since $1 - 1 = 0$. And only

$$P_f(-1) = P_{\min}(-1)$$

remains, where $P_{\min} = 1 + 2t + t^2$. Of course, this implies that $P_f(-1) = 0$ and does not give any useful information about the number of critical points. These

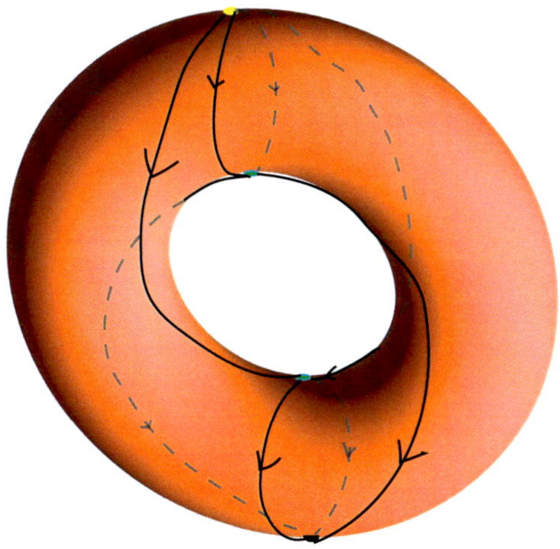

FIGURE 14. In the generic torus case there are at least four critical points. The picture shows a "minimal" configuration with four critical points and all rigid flow lines.

are precisely the type of expressions we derived previously for the various pictures. A better estimate for the number of critical points is given using the evaluation at $t = 1$

$$P_f(1) = P_{\min}(1) + 2 \cdot Q_f(1) = 4 + 2 \cdot Q_f(1) \geq 4,$$

where we note that $P_f(1)$ is the total number of critical points for f. We deduce that we must have at least four critical points.

Our Morse equality even tells us more. To see this, we partition the set of critical points of f into the set of local minima, passes, and local maxima (indexed by 0, 1 and 2, respectively)

$$\mathrm{Crit}(f) = \mathrm{Crit}_0(f) \sqcup \mathrm{Crit}_1(f) \sqcup \mathrm{Crit}_2(f).$$

As such, we can write $P_f(t) = \sum_{i=0}^{2} \sharp\,\mathrm{Crit}_i(f) \cdot t^i$. Then (6) can be rewritten as

$$\sum_{i=0}^{2} \sharp\,\mathrm{Crit}_i(f) \cdot t^i = 1 + 2t + t^2 + (1+t) \cdot Q_f(t) \tag{7}$$

and we deduce the inequalities[17]

$$\sharp\,\mathrm{Crit}_0(f) \geq 1, \ \sharp\,\mathrm{Crit}_1(f) \geq 2 \ \text{ and } \ \sharp\,\mathrm{Crit}_2(f) \geq 1.$$

In our present torus case we obtain that there is at least one global minimum, two saddle points and one global maximum.

1.3. The Big Picture

Looking at the Figures 12, 13 and 14 and recording the data, we have seen an interaction between calculus and topology. In fact, what we mean by topology is "algebraic topology," a field whose goal is to describe properties of a large class of spaces by means of algebra.

One of the first instances of studying such a relationship is Eulers's polyhedral formula $V + F - E = 2$ for convex 3-dimensional polyhedra (V, F and E are the numbers of vertices, edges and faces). In any given case the left hand side of the formula depends on the (local) construction of the convex polyhedron and the right-hand side captures the global nature of the polyhedron by the number 2. If the polyhedron is not assumed to be convex we might obtain different integers on the right hand side, which again reflect global topological properties.

The reader should notice the similarity to our discussion leading up to the Morse equalities. In the simple case discussed, we associated to a smooth function f defined on Λ the Poincaré polynomial $P_f(t)$ which keeps track of the critical points. P_f can be written in the form $P_\Lambda(t)+(1+t)Q_f(t)$, where $P_\Lambda(t)$ is a quantity reflecting properties of the space Λ, and $Q_f(t)$ records the effects of wiggling in the case that f has more critical points than Λ demands. Even if a smooth function f on a smooth space Λ is an imperfect choice due to the nuisance contributions of the birth-death mechanism, it still can be used to construct an invariant of the space. To see this, just rewrite the Morse equality:

$$P_\Lambda(t) = P_f(t) - (1+t)Q_f(t). \tag{8}$$

So far we considered very simple spaces equipped with a version of calculus so that it makes sense to talk about the derivative of a function. The critical points of a function correspond to valleys and peaks, and the algebraic description—using polynomials—gives a relationship between various quantities. To make such structural statements the birth-death mechanism is crucial.

Mathematicians have developed general methods to define versions of multivariable calculus on suitable spaces, [**MH, Tu, Lee, AbM**][18]—the fields of differential geometry and differential topology. Spaces equipped with a version of multivariable calculus are called smooth manifolds. Of course, the construction of smooth spaces comes with a number of definitions and results which describe salient features. The philosophy is not too difficult to understand: The reader will have an intuitive understanding about a smooth curve in the plane or in the three-dimensional Euclidean space, see Figure 15. Similarly imagining a smooth curve or smooth surface in the three-dimensional space should be no problem. It can be shown that a smooth function on such a surface or curve is nothing else but the restriction of a smooth map on the ambient Euclidean space—a smooth map as we know it from calculus.

FIGURE 15. A curve and a two-dimensional surface embedded in three-dimensional Euclidean space.

We are interested in smooth manifolds, or more colloquially smooth spaces of higher dimensions. These spaces are equipped with a version of multivariable calculus and we can talk about smooth maps defined on them. It is known by the Whitney embedding theorem that smooth manifolds can be viewed as lying in a high-dimensional Euclidean space so that smooth maps on them are again restrictions of smooth maps on the ambient space. This provides us with a framework to talk about smooth functions and take derivatives—denoted by $df(x)$ generalizing

$f'(x)$. A critical point for f is a point where $df(x) = 0$—minima and maxima are critical points. At critical points, we can check the so-called Hessian $\mathrm{Hess}(f, x)$ which is a substitute for the second derivative $f''(x)$ in this more general context. The Hessian also gives us a way to define the Morse index in higher dimensions. The key points which we have discussed in the one- and two-dimensional cases remain true in every finite dimension: Morse theory generalizes to the higher-dimensional cases.

Given a smooth map f on a smooth manifold Λ, where we assume a nice behavior near the critical points (well-formedness), there is a relationship between the polynomial data $P_f(t)$ and a polynomial $P_\Lambda(t)$ in the usual form

$$P_f(t) = P_\Lambda(t) + (1+t)Q_f(t), \tag{9}$$

where Q_f is a polynomial with non-negative integer coefficients accounting for the noise from the birth-death mechanism.

There is a lot one can say about P_Λ. We have already described it as a count of critical points which at a minimum must be there. Mathematicians have studied this question further and it is closely related to the fact that spaces can have holes and these holes can be counted—in fact, P_Λ is a hole count.[19]

Spaces equipped with a version of calculus have a dimension $\dim(\Lambda)$—just as the Euclidean spaces have—and the takeaway from (9) is then:

For every integer $k = 0, \ldots, \dim(\Lambda)$ the number of critical points of Morse index k is bounded from below by a count of k-dimensional holes.[20]

The mathematical development of Morse theory leading to these types of results follows along the lines envisioned by Marston Morse, himself about one-hundred years ago. Here is an excerpt from the introduction of Morse's highly influential book, [**Mo2**]:

"For several years the research of the writer has been oriented by a conception of what might be termed macro-analysis. It seems probable to the author that many of the objectively important problems in mathematical physics, geometry, and analysis cannot be solved without radical additions to the methods of what is now strictly regarded as pure analysis. Any problem which is nonlinear in character, which involves more than one coordinate system or more than one variable, or whose structure is initially defined in the large, is likely to require considerations of topology and group theory in order to arrive at its meaning and its solution. In the solution of such problems classical analysis will frequently appear as an instrument in the small, integrated over the whole problem with the aid of group theory or topology. Such conceptions are not due to the author. It will be sufficient to say that Henri Poincaré was among the first to have a conscious theory of macro-analysis, and of all mathematicians [he] was doubtless the one who most effectively put such a theory into practice."

In the first occurrence of what are called the "Morse inequalities," [**Mo1**], which later led to (9), Morse relied on a mathematical concept called homology numbers. Homology, a concept in algebraic topology, was initially developed in the 19th century by Bernhard Riemann and Enrico Betti. They introduced what are called "Betti numbers." These numbers provided a way to classify so-called topological spaces. In fact, this classification is nothing else but a count of holes of various dimensions. Counting of holes requires precise mathematical definitions and, in fact, a whole theory. In the early 20th century, Henri Poincaré extended this work by formalizing the concept of homology in higher dimensions.

FIGURE 16. Emmy Noether.

However, it was Emmy Noether who, around 1925, revolutionized the understanding of homology by introducing the idea of homology groups. She recognized that homology could be better understood as a group structure—a relationship to symmetries—rather than just numerical invariants like Betti numbers. This shift allowed for a more abstract and general framework, connecting homology to algebraic structures.

Mathematicians define the "hole structure" of a space by analyzing the structure of its associated homology groups—they apply the tools from the highly-developed field of algebraic topology.

Crossing into infinite dimensions is necessary in order to make Morse theory applicable to a wider range of problems. When we view the finite-dimensional theory as the study of the critical points of a smooth function on an abstract finite-dimensional manifold, then the infinite-dimensional version representing the same generality would be the study of a function on what are called Hilbert or Banach manifolds. A generalization is accomplished by Richard Palais and Steven Smale, [**PS**].

FIGURE 17. Richard Palais (left) and Stephen Smale.

This generalization can be viewed as the limit of finite-dimensional situations as the dimensions increase to infinity. In this case we can define the Morse indices as before, and the Palais-Smale generalization establishes a relationship between critical points of Morse index k and the count of k-dimensional holes for every integer $k = 0, 1, \ldots$. Everything we know from finite dimensions generalizes beautifully under a minimum of assumptions—under the so-called Palais-Smale condition (PS). See the Appendix on page 125 for details.

However, there were plenty of smooth maps on infinite dimensional spaces which did not have a finite Morse index and could not be studied by the Palais-Smale extension. Maybe the ideas could be generalized. This raised the question of whether in some way one could also study smooth maps on infinite-dimensional spaces that only had infinite Morse indices. Perhaps there was an ingenious way relating the number of critical points with infinite Morse indices to the count of infinite-dimensional holes?

The problem is that the classical machinery of algebraic topology will not find any infinite-dimensional hole since there is too much room in infinite dimensions. Alas, this possibility was quickly dismissed. And for a good number of reasons as articulated by Jürgen Moser in the case of the action functional, which we shall introduce in the next chapter. We shall describe his pessimistic view of the situation in Chapter 3—but we can already describe the seemingly ugly truth that motivated Moser's opinion:

There is no such thing as an infinite-dimensional hole in an infinite-dimensional space![21]

This is a punchline of sorts. It seems hopeless—there is no obvious relationship between the critical points of infinite Morse index and the topology of the ambient space.

CHAPTER 2

The (Bad) Action Functional

Regarding the aforementioned class of action functionals that for the most part was considered useless: Some of these principles are associated with a particular class of dynamical systems called Hamiltonian systems and play an important role in physics. They characterize the dynamics of a mechanical system as a critical point of some map. Philosophically speaking, of all the things which could happen, what actually happens is characterized by being a critical point of some quantity. This is in line with a modification of Euler's remark "...nothing at all takes place in the universe in which some rule of maximum or minimum does not appear," if we allow besides minima and maxima other types of critical points. In fact, this broader view had emerged in physics and later became the starting point for quantum mechanics. This action principle inspired the path integral formulation of quantum mechanics, developed by Richard Feynman.

The theory of dynamical systems was born out of celestial mechanics, the study of the movement of the stars, the planets, and their moons. An observational science, celestial mechanics advanced through investigations by the likes of Tycho Brahe and Johannes Kepler, Isaac Newton and Gottfried Wilhelm Leibniz, Leonard Euler, Joseph-Louis Lagrange, Rowan Hamilton, and Henri Poincaré.

Newton's genius, of course, was demonstrated by his fundamental idea that the gravitational force experienced by a falling apple on Earth should be the same force experienced between two celestial bodies. His calculus made it possible to write down a set of differential equations describing the movement of celestial bodies under the force of gravity. He also showed that Kepler's observations—and approximate laws—would follow from the same equations.[22] For example, Newton considered the equation for the Sun-Earth-Moon system—a complicated system for which one cannot write solutions in an explicit form. Historically, and surprisingly, the study of these equations has often served as a benchmark for testing new mathematical ideas, even in modern times.[23] While Newton's mechanics is based on the idea of acting forces ($F = ma$), analytical mechanics, most prominently the Lagrangian and Hamiltonian versions, is rooted in the concept of energy—kinetic and potential energy.[24]

The differential equations in celestial mechanics—which can be written either as a second-order equation in Lagrangian mechanics, or as a first-order system in Hamiltonian mechanics—are formally obtained from variational principles. The dynamics of Lagrangian mechanics is governed by the principle of the least action (or the extended principle of stationary action). Hamiltonian mechanics turns on the conservation of energy, as well as an action principle—the so-called Hamiltonian action principle.

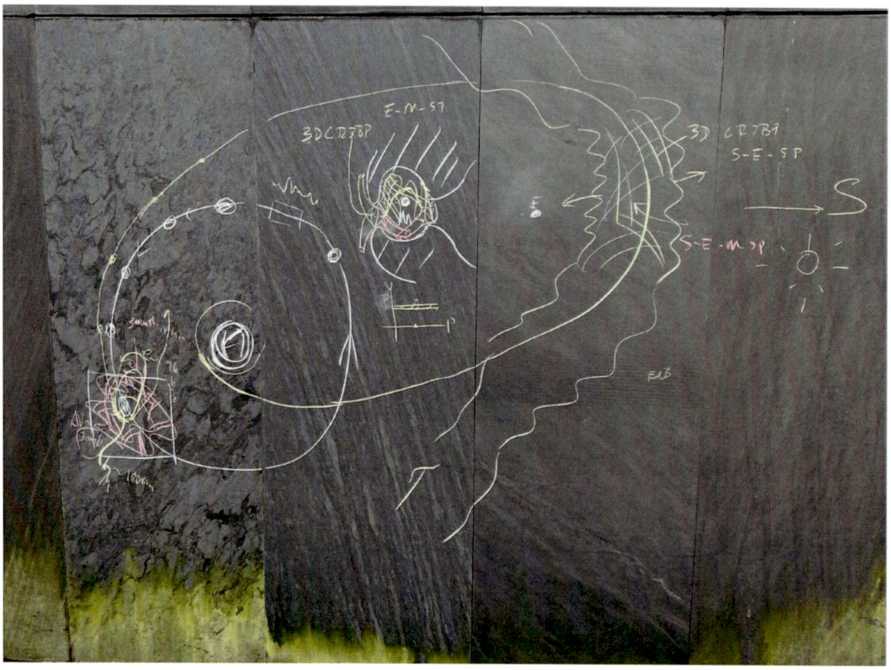

FIGURE 1. Outdoor blackboard at IAS (2020) with Ed Belbruno's description of the ballistic capture applied to the Japanese lunar probe Hiten.

Action principles are variational principles. Whereas the least action principle is in most cases mathematically useful, the Hamiltonian action principle was considered useless as a variational principle for mathematical purposes. However, as already stated, it was conceptually useful from a physics perspective as a starting point for the description of quantum mechanics.

As it turns out, many of the systems arising in physics exhibit similar features. Over time, this led to the development of Hamiltonian dynamics as a subfield of the theory of dynamical systems. Today such Hamiltonian systems are relevant in physics, with applications ranging from the beam physics of particle accelerators to the design of scientific space missions.[25]

While Poincaré expected that many Hamiltonian systems should typically show chaotic behavior, it could never be proved—for good reason. It only became clear with the development of KAM-theory, due to Kolmogorov, Arnold and Moser, [**Du**]. KAM-theory showed that there is more stability in such systems than Poincaré expected. This coexistence of stable and random motions lies behind the interesting properties exhibited by Hamiltonian systems, [**Mos0**].

Poincaré, described as "The Last Universalist" [**GG**], was an engineer, mathematician, theoretical physicist and a philosopher of science—he was a leader in whatever he undertook. His comprehensive and integrative views led to innovative ideas in mathematics and physics. He promoted the qualitative understanding of systems—for instance studying the properties of solutions, even if it wasn't possible to explicitly solve the equations.[26] Ideas arising from his work not only led to advancements in dynamical systems (particularly the qualitative theory of which he is considered the founder), but also to some ideas[27] which are of a symplectic nature and lead to the dramatic development of symplectic geometry in the 1980s.[28] While these two spheres of ideas—dynamical and symplectic—emanated from Poincaré's integrated point of view, the historical development separated these ideas and for a long time there was not much interaction.

Today, having powerful computers at our disposal, a deep qualitative understanding of the mathematical phenomena is important for analyzing the stability of numerical schemes since it is generally impossible to solve the underlying equations explicitly. The numerical schemes are only approximations and errors accumulate, particularly when studying systems that describe unstable chaotic phenomena. For example, when studying the dynamics of our solar system by computer, with precise modeling even incorporating general relativity, it is impossible to predict where Pluto will be in, say, 80 millions years.[29] The chaotic behavior of our solar system—chaos in slow motion—requires long-term computations and massive amounts of memory. Even if the entire universe was transformed into a computer (with the type of architecture currently used) there would not be enough memory to compute the Pluto prediction with a sufficient accuracy, for the simple reason that there are not enough electrons in the (observable) universe.[30]

Our solar system consists of all the bodies which are gravitationally bound to the sun. The main player which (approximately) holds everything together is the sun. It is followed by the planets, then larger celestial bodies like the "deplanetized" Pluto and on the lower end by space junk. All the players interact by the gravitational forces in a way, where the bigger bodies have an outsized impact on the smaller ones. From this perspective Pluto is the "plaything" of the sun and the eight planets. The ever changing ebb and tide of gravitational forces experienced by Pluto creates a chaotic system. On a human time scale it is chaos in slow motion.

In order to make predictions astronomers build a mathematical model. These are quite complicated since relativity theory has to be built in for accuracy. Mathematicians call such models a dynamical system and we shall describe the idea in a visual way in the next section.

2.1. A Visual of Dynamical Systems—The Flow

We start with a physical system for example the sun and the eight planets. Simplifying matters we view the celestial bodies as mathematical points. We fix a coordinate system and describe their locations, three coordinates, with respect to this system. Nine bodies produce $9 \times 3 = 27$ numbers. Each point has a velocity which is directed in space and is described by three numbers. Incorporating the different masses we replace the velocity by a better physical quantity, namely the momentum of the point. Ultimately our system is described by 54 numbers which we can view as a point in a 54-dimensional space Λ. When the planets and the sun move, these numbers constantly change and the evolution of the system can be described by a moving point in a 54-dimensional space. How this point moves in the space is described by a differential equation which is defined on this space. Such an equation only makes sense if a form of calculus exists on our 54-dimensional space and it does—Λ is a smooth space away from a singular set describing two or more bodies crashing into each other. We shall ignore subtleties like singularities for our presentation. We can visualize the mathematical modeling as constructing a smooth space—its points describe the states of the system—together with a prescription implementing physical laws, which govern how points move in time.

Hence the following picture comes into mind, see Figure 2.

FIGURE 2. The smooth space Λ and a state $x \in \Lambda$.

Over time the states will change according to some physical law and this change can be described as follows: If we have the state x now, then in time t it will change to a state denoted by $t \star x$. Since there is no change in time 0 we obviously obtain our first rule:

$$0 \star x = x.$$

Another requirement is that if we have two times s and t then the evolution from x to $(s+t) \star x$ should be the same as having x first evolve to $t \star x$ and then $t \star x$ to $s \star (t \star x)$ (and similarly $x \to s \star x \to t \star (s \star x)$). This gives the second rule:

$$(s+t) \star x = s \star (t \star x).$$

This leads to the important definition of a flow. A flow on a space Λ is given by a map which associates to a time t and a state $x \in \Lambda$ a new state $t \star x$ subject to two rules previously described.

In general, given a state x, we obtain the map

$$\mathbb{R} \to \Lambda : t \to t \star x$$

which can be viewed as a path in Λ documenting the history of the state x, i.e. $t \star x$, where $t < 0$, and the future of x, i.e. $t \star x$ for $t > 0$. This path is called a flow line. This is shown in Figure 3.

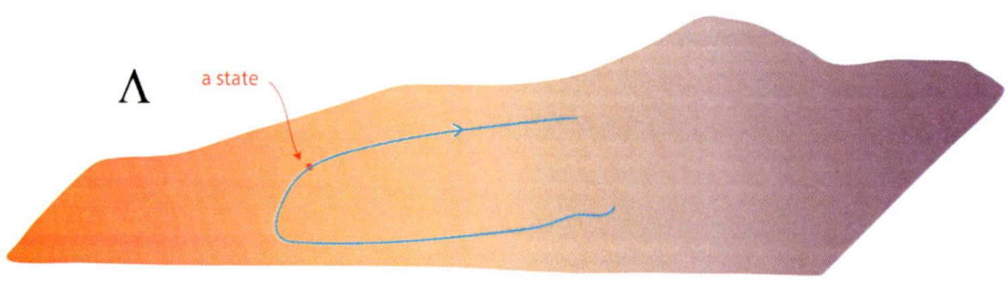

FIGURE 3. The history and future of a state x, i.e. the path $\mathbb{R} \to \Lambda : t \to t \star x$.

A flow is a general concept with many applications and we used this idea to illustrate the idea of a dynamical system. There is also the important notion of a semi flow, where $t \star x$ only is defined in forward time, i.e. $t \geq 0$—this is an abstraction covering processes which are not reversible.

One can distinguish different classes of flows specifying additional properties. An important class in our context is that of a Hamiltonian system. These systems play an important role in physics and appear in real life situations. Hamiltonian systems have remarkable properties since in some sense such a system performs a balancing act by exhibiting stable behavior as well as unstable chaotic behavior. In the next section we lay the ground work describing the structures which make Hamiltonian systems so interesting.

2.2. A Function on the Space of Loops

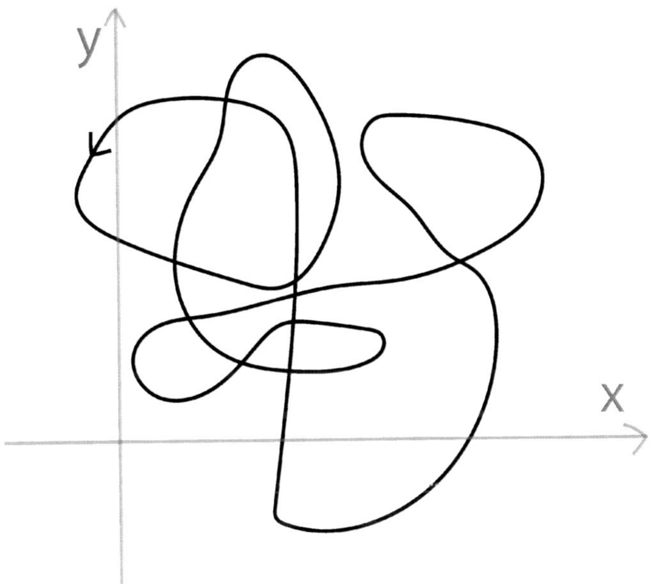

FIGURE 4. The curve is oriented by the arrow; and if it intersects itself, it does so nicely.

A good starting point to introduce the action functional is the notion of the (Hamiltonian) action of a loop in the space $\mathbb{R}^{2n}$. First, in dimension two, we consider a sufficiently nice closed curve in the plane $\mathbb{R}^2$. It is given by a map

$$z : S^1 \to \mathbb{R}^2,$$

see Figure 4. Here $S^1 = \mathbb{R}/\mathbb{Z}$ is the "circle" obtained by identifying two points on the real line provided their difference is an integer.[31] Sufficiently nice means that self-intersections are transversal. The goal is to associate a real number to any curve of the type shown in Figure 4.

Figure 5 describes an algorithm to compute a real number called the action associated to the loop shown in Figure 4. We start with the given loop and then we pick a segment like the one colored red which closes up. It encloses the area $A > 0$. Since the curve encloses the area counter clockwise we record the number $+A$ and delete the segment to obtain a new curve. Then we pick a second segment, which in our case encloses the area B, but clockwise, and we record $-B$. We delete the segment, obtain a new loop and the recorded number $A - B$. After a finite number

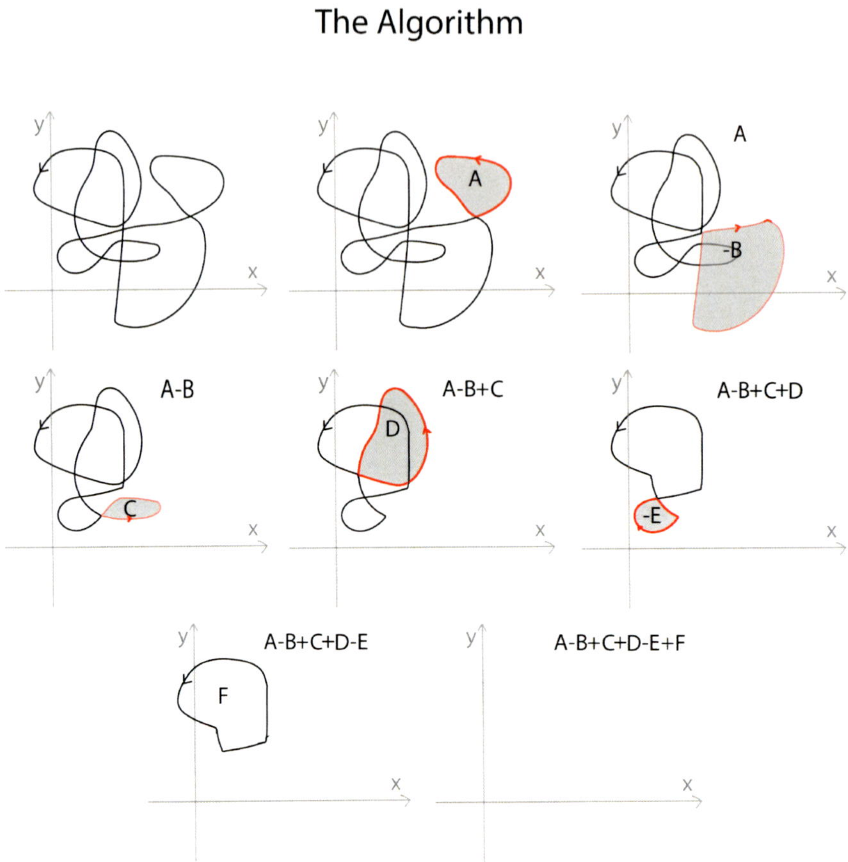

FIGURE 5. A more general curve and an algorithm.

of steps we obtain a real number and no curve is left. One can show that the final number does not depend on the choices which were made.

This procedure associates to a loop in the plane $\mathbb{R}^2$, say $z : S^1 \to \mathbb{R}^2$ a number $\mathcal{A}_{\mathbb{R}^2}(z)$, called the action of the loop z. Denoting the collection of such loops by $\text{Loop}(\mathbb{R}^2)$ the procedure will construct a map

$$\mathcal{A}_{\mathbb{R}^2} : \text{Loop}(\mathbb{R}^2) \to \mathbb{R} : z \to \mathcal{A}_{\mathbb{R}^2}(z).$$

It can be shown that even for not so nice curves one can define the action.[32]

We can generalize this to higher even dimensions as follows. View the Euclidean space of dimension $2n$, say $\mathbb{R}^{2n}$, as the product of n-many two-planes

$$\mathbb{R}^{2n} = \mathbb{R}^2 \times \cdots \times \mathbb{R}^2.$$

Now given a loop in $\mathbb{R}^{2n}$, say $z : S^1 \to \mathbb{R}^{2n}$, we can write

$$z(t) = (z_1(t), \cdots, z_n(t)),$$

where every $z_i : S^1 \to \mathbb{R}^2$ is a loop in a plane $\mathbb{R}^2$. We already know how to compute the action of each z_i and we define

$$\mathcal{A}_{\mathbb{R}^{2n}} : \text{Loop}(\mathbb{R}^{2n}) \to \mathbb{R}$$

by taking the sum of the actions of the loops z_i

$$\mathcal{A}_{\mathbb{R}^{2n}}(z) = \mathcal{A}_{\mathbb{R}^2}(z_1) + \ldots + \mathcal{A}_{\mathbb{R}^2}(z_n). \tag{10}$$

In other words we project z onto the n-many planes, obtaining loops in $\mathbb{R}^2$, computing their actions, and summing them up. We shall call $\mathcal{A}_{\mathbb{R}^{2n}}(z)$ the action of the loop $z \in \text{Loop}(\mathbb{R}^{2n})$.

We have given an intuitive definition of $\mathcal{A}_{\mathbb{R}^{2n}}$ which shows that two-dimensional areas, the orientations of the loops, as well as certain projection onto two-dimensional planes, play a role. As an aside, one can give a precise mathematical formula for the action using some calculus ideas.[33]

Before we use the function $\mathcal{A}_{\mathbb{R}^{2n}}$ we should get acquainted with what is called generalized mechanics and begin with a short historical detour.

2.3. Hamiltonian Dynamical Systems

In the Principia Mathematica, Isaac Newton formulated three laws, governing mechanical systems:

(1) An object will not change its motion unless a force acts on it.
(2) The force on an object is equal to its mass times its acceleration.
(3) When two objects interact, they apply forces to each other of equal magnitude and opposite direction.

Using calculus, developed by Newton and independently by Leibniz, one could write down equations describing mechanical systems and sometimes even solve them explicitly. The basic concepts were: force, acceleration and mass, position (in configuration space) and velocity. In principle Newton's insight made it possible to describe the future motion of a mechanical system based on its known present position and velocity. To illustrate this point, consider the movement of a satellite in the Earth-Moon system. There are different ways to use the Newtonian ideas to derive a mathematical model but all descriptions would produce, with respect to a suitable three-dimensional coordinate system, the path of the satellite and its velocity at any given time. The state of the system is given by six numbers. The first three describe the position in the coordinate system—or more generally the configuration space; the second three numbers describe the velocity vector. Of course, in this example we are given six numbers which change in time and we have a strict distinction between position and velocity.

Other viewpoints were developed leading to what is now called generalized mechanics based on the new concepts of energy and momentum. In this context one considers the position of the satellite as a point in the so-called phase space. In our example this would be a six-dimensional space describing in an aggregate way a generalized position from which the predictions of the Newtonian models could be derived. These new ideas, which ultimately were important for the development of quantum mechanics, came with a price: The mathematical set-up becomes more abstract and harder to grasp on an intutitive level. Now we consider the $2n$-dimensional Euclidean space $\mathbb{R}^{2n}$ equipped with an additional piece of data ω, called a symplectic form, the prototype of a phase space in generalized or analytical mechanics.

Here is the salient point: The piece of data ω in the generalized mechanics model—the one which causes most headaches—is equivalent to $\mathcal{A}_{\mathbb{R}^{2n}}$ and instead of $(\mathbb{R}^{2n}, \omega)$ one can consider $(\mathbb{R}^{2n}, \mathcal{A}_{\mathbb{R}^{2n}})$. Hence instead of viewing points in $\mathbb{R}^{2n}$

we think of $\mathbb{R}^{2n}$ as being occupied by loops $z : S^1 \to \mathbb{R}^{2n}$, that is loops of states $S^1 \ni t \to z(t) \in \mathbb{R}^{2n}$ equipped with a function which associates to the loop the action $\mathcal{A}_{\mathbb{R}^{2n}}(z)$.

$(\mathbb{R}^{2n}, \mathcal{A}_{\mathbb{R}^{2n}})$ is a local model. It is possible to patch local models together to create more complicated phase spaces. This is to the smooth surface case, where more complicated surfaces are being patched together from the local calculus models $\mathbb{R}^{2n}$.

2.3.1. Predictions and the Cauchy Problem. Taken alone $(\mathbb{R}^{2n}, \mathcal{A}_{\mathbb{R}^{2n}})$ is just a phase space with some structural data $\mathcal{A}_{\mathbb{R}^{2n}}$, but there is no physics yet! To change this we have to allocate to every state $z \in \mathbb{R}^{2n}$ an energy $H(z)$. This means we need a smooth map $H : \mathbb{R}^{2n} \to \mathbb{R}$. In praxis H reflects the applicable physical laws for the system which is being described. One can also consider time-dependent energies $H : \mathbb{R} \times \mathbb{R}^{2n} \to \mathbb{R} : (t, z) \to H(t, z)$ if required by the physical modeling. To $(\mathbb{R}^{2n}, \mathcal{A}_{\mathbb{R}^{2n}}, H)$ we can associate a system of ordinary differential equations which can be written as

$$\dot{q}_i = \frac{\partial H}{\partial p_i}(t, z) \text{ and } \dot{p}_i = -\frac{\partial H}{\partial q_i}(t, z) \text{ for } i = 1, \ldots, n. \tag{11}$$

This type of differential equation is called a Hamiltonian system. Mathematicians for a long time have developed tools to study some aspects of such equations. The upshot is the predictive power: If we currently witness the physical system in the state $z_0 \in \mathbb{R}^{2n}$, then solving the ordinary differential equation with the initial data z_0 gives us a curve $t \to z(t) \in \mathbb{R}^{2n}$ through the space of states with the property $z(0) = z_0$, where $z(t)$ is the state the system will be in at time $t > 0$ or was at $t < 0$. The ordinary differential equation (11) together with the prescribed initial data z_0 is called a Cauchy problem and such problems can be solved numerically quite efficiently—with limitations (lack of electrons) as the impossibility of the prediction of Pluto's long-term orbit shows, recall the discussion on page 21.

There is an aspect one should be aware of, since the interesting behavior of Hamiltonian systems is a consequence. When we consider the ordinary differential equation (11) we can do the following. Consider a loop $Z : S^1 \to \mathbb{R}^{2n} : s \to Z(s)$ which by our previous discussion has an action denoted by $\mathcal{A}_{\mathbb{R}^{2n}}(Z)$. Next we consider for every $s \in S^1$ the point $Z(s)$ as the initial data for (11) which after some $T \in \mathbb{R}$ gives us the new state $T \star Z(s)$. Doing this for all $s \in S^1$ we obtain the new loop $s \to T \star Z(s)$, which we shall denote by Z'. The miracle is the equality

$$\mathcal{A}_{\mathbb{R}^{2n}}(Z') = \mathcal{A}_{\mathbb{R}^{2n}}(Z).$$

In other words, the Cauchy problem preserves the action of a loop of initial data. This is not too difficult to prove, but the fact that ancient scientists involved in celestial mechanics figured out that the motion of sun and planets can be described by a system which has this kind of property is remarkable and turned Hamiltonian systems into an important class of dynamical systems.

2.3.2. Cauchy Problem vs. Special Solutions. The Cauchy problem has to be contrasted with the task of finding special solutions with prescribed properties. For example a scientific space mission relies on particular orbits with very special properties—addressing the intended mission goals, time, fuel, and engineering constraints. Finding special orbits is like finding a needle in a haystack in large part due to the fact that Hamiltonian systems very often show a chaotic behavior and in real life, like mission design, engineering constraints make it even more difficult. The search for special orbits is a complicated task and in general not well-understood.[34] However, tremendous progress has been made in developing the mathematical tools for such tasks.

One of the easiest types of special orbits are rest points and periodic orbits. Given a smooth function $H : \mathbb{R}^{2n} \to \mathbb{R}$, the energy distribution or Hamiltonian, we can define a map which associates to a loop $z : S^1 \to \mathbb{R}^{2n}$ the average H-energy along it as the number $\int_0^1 H(z(t))dt$. This defines a map

$$z \to \int_0^1 H(z(t))dt$$

which allows us to built the perturbed action $\mathcal{A}^H_{\mathbb{R}^{2n}}$ defined by

$$z \to \mathcal{A}_{\mathbb{R}^{2n}}(z) - \int_0^1 H(z(t))dt \tag{12}$$

on the space Loop($\mathbb{R}^{2n}$) of all loops in $\mathbb{R}^{2n}$. Formally the critical points of $\mathcal{A}^H_{\mathbb{R}^{2n}}$ are 1-periodic orbits of the Hamiltonian system associated to H.

Poincaré was convinced that the collection of all periodic orbits in many low-dimensional Hamiltonian systems knows a lot about the overall dynamics. He did not have the tools to prove this, but results obtained in the last two decades confirm his belief. Today periodic orbits play an important role in the theory of symplectic invariants. But how does one find them? Until 1978, the perturbed action functional in (12) was considered completely useless for rather convincing reasons, and this will be part of our story which leads to Floer theory.

A rigorous treatment of $\mathcal{A}^H_{\mathbb{R}^{2n}}$ or its variants on more general phase spaces under a host of conditions on H is now possible. This is part of our story. Finding specific types of interesting orbits for Hamiltonian systems can very often be formulated as finding a suitable critical point of the (perturbed) action functional. One can study the existence of solutions satisfying other suitable boundary conditions, which also lead to variational problems similar to those we have just discussed.

2.3.3. The General View. One of the features of Hamiltonian dynamics is that many questions about the systems can be formulated as geometric problems. The appropriate mathematical language is symplectic geometry, developed by Vladimir Arnold and Alan Weinstein in the 1960s and 1970s.

"Symplectic," from the Greek symplektikos, means "intertwining."[35] Mathematically speaking, symplectic geometry is the study of pairs (M, ω), where M is a smooth manifold of even dimension and ω is a non-degenerate closed two-form. This extends the ideas introduced around the local model $(\mathbb{R}^{2n}, \omega)$. As we already mentioned we can view (M, ω) as patched together from multiple copies of the local model. As a consequence, instead of considering (M, ω), we could think of M as equipped with a structure, having certain properties, which associates to small loops a real number, similar to what we have constructed on $\mathbb{R}^{2n}$.[36]

As it turns out we can define in general for a given (M, ω) an action functional for small loops. In some cases, knowing the action for small loops one can define an action for large (contractible) loops as well. This works for the standard torus with the standard symplectic form, but not for the two-sphere equipped with a symplectic form.

FIGURE 6. "Symplectic geometry?!"[37]

This general viewpoint—instead of (M, ω) consider the loop space and the action functional—starts with the seminal work of Conley and Zehnder, [**CZ**].

The standard symplectic torus (T^{2n}, ω) is replaced by the action

$$\mathcal{A}_{T^{2n}} : \mathrm{Loops}_{\mathrm{contr}}(T^{2n}) \to \mathbb{R}$$

associating to a contractible loop a real number and the original torus appears as the smooth manifold of critical points. This turns out to be a powerful equivalent point of view which, however, does not come entirely out of the blue and the background story is told in the next chapter.

2.4. The Arnold Conjecture

In the sixties Arnold had formulated several conjectures in symplectic geometry, generally referred to as the Arnold conjectures. We finish this section by describing one of them.

Given a closed smooth manifold M and a smooth function

$$f : M \to \mathbb{R}$$

we can count its critical points. The count, denoted by $\sharp f$, results in either a non-negative integer or ∞. We also have the subclass of functions which are Morse, i.e. the critical points are non-degenerate.

Now we define two numbers in $\{0, 1, \cdots\}$ as follows:

$$\begin{aligned} \sharp_{\mathrm{Morse}}(M) &:= \min\,\{\sharp f \mid f \text{ is Morse}\} \\ \sharp(M) &:= \min\{\sharp f \mid f : M \to \mathbb{R} \text{ is smooth}\}. \end{aligned} \tag{13}$$

The first number is the minimal number of critical points a Morse function on M has to have and as we already know Morse theory is able to give lower bounds for this number. The second number is generally smaller by allowing arbitrary smooth maps without requiring non-degeneracy of the critical points. If the dimension of M is at least two it follows that any function f has a global minimum and a global maximum both obtained at two different points. Hence, if $\dim(\Lambda) \geq 2$ we already know that $\sharp_{\mathrm{Morse}}(M) \geq \sharp(M) \geq 2$.

We assume that (M, ω) is a closed symplectic manifold. In this case one can define an action $\mathcal{A}_M(z)$ for sufficiently small loops. For arbitrarily large but contractible loops there are ambiguities, and one can only define a multi-valued function.[38]

Given a smooth time-dependent Hamiltonian $H : S^1 \times M \to \mathbb{R}$ we obtain a time-dependent vector field X_{H_t} on M and consider the ordinary differential equation

$$\dot{z} = X_{H_t}(z),$$

which in so-called local coordinates looks precisely as we described it in the $\mathbb{R}^{2n}$-case. One is interested in the number of 1-periodic solutions of this ordinary differential equation and one would like to count them. The periodic solutions can, at least formally, be found as critical points of a multivalued functional. That the functional is multivalued complicates matters. One can impose a non-degeneracy condition on H which turns out to be equivalent to the condition that the critical points of the multivalued functional are non-degenerate. In this context Arnold made the following conjecture.

ARNOLD CONJECTURE:. Consider the closed symplectic manifold (M, ω) equipped with 1-periodically forced Hamiltonian $H : S^1 \times M \to \mathbb{R}$. Then the associated Hamiltonian system has at least $\sharp(M)$-many periodic orbits. and if H is non-degenerate, it has at least $\sharp_{\text{Morse}}(M)$ many periodic orbits.

In the case of the standard torus (T^{2n}, ω) the lower bounds predicted by Arnold are $2n + 1$ and in the non-degenerate case 2^{2n}. This is what Charley Conley and Eduard Zehnder proved in their breakthrough result, [**CZ**].

The reader will find some of the basic mathematical background in the books [**AbM**, **Ar1**]. Arnold's motivation for his conjecture comes from the following fact. With some work one can show that the fixed points of a Hamiltonian diffeomorphism, which is sufficiently small, correspond one-to-one to the critical points of a function. In this case one can use Morse theory, or the so-called Liusternik-Shnirelman theory for the function to estimate the number of fixed points. The Arnold conjecture essentially stipulates that the smallness assumption is not needed—quite a daring conjecture.[39]

CHAPTER 3

Moser's Error and Beyond

FIGURE 1. Jürgen Moser.

In a paper published in 1976 Jürgen Moser, [**Mos**], gives a rather convincing argument as to why the action functional cannot be used for a global existence proof. One might expect that his opinion could safely be taken as hardcore and correct. After all, Moser was "the Pope" of dynamical systems. And outside of the mathematical realm, he had a reputation as something of a tough guy.[40] There was a telling incident in New York City, in the 1970s. Walking down 6th Avenue, Moser, then a professor at the Courant Institute, watched with disgust as a passing car ejected garbage onto the street. When the car stopped at a red light, its passenger-side window still open, he picked up the trash, approached the car, said " Gentlemen, you lost something" and stuffed the trash back through the window.[41]

Moser's reasoning was based on an educated understanding of Morse theory: Having critical points with both an infinite Morse index and an infinite Morse co-index destroys the relationship between the global topology and the local structure near the critical points, which is at the core of finite-dimensional Morse theory or using a previously described insight: There are no infinite-dimensional holes in infinite-dimensional (Hilbert) manifolds; there is just too much room in an infinite-dimensional space, see Appendix F on page 129.

The back story is as follows: In a paper [**We3**] published in 1973, Alan Weinstein studied local problems in Hamiltonian mechanics using the modern language he had developed. These results triggered Moser's paper, "Periodic orbits near an equilibrium and a theorem by Alan Weinstein," [**Mos**], which provided a generalization. Moser was interested in finding periodic orbits of a Hamiltonian system near an equilibrium point. He explains the mathematical problems with this principle and he also outlines how in his local considerations he will be able to overcome these difficulties. From a historical perspective his statement given in [**Mos**] is important. He writes: "However, this variational principle is very degenerate, for example even the Legendre condition is violated, and is certainly not suitable for an existence proof."[42]

One could try to base the solution of this boundary value problem on the well known variational principle which calls for the extrema of the functional

$$S = \int \sum y_k\, dx_k - \lambda(H-c)\, ds = \int_0^1 \left(\left(y, \frac{dx}{ds}\right) - \lambda(H-c)\right) ds \tag{1.5}$$

over the class of periodic vector functions $z(s) = (x(s), y(s))$ of period 1 and the parameters λ. The Euler equations for this variational principle are given precisely by (1.2) together with $H = c$.

However, this variational principle is very degenerate, for example even the Legendre condition is violated, and is certainly not suitable for an existence proof. To overcome this difficulty we solve the above boundary

FIGURE 2. Moser's "philosophical" error in [**Mos**].

Moser is talking about the Hamiltonian variational principle which is much more general than the least (or even the stationary) action principle for the Lagrangian formulation of mechanics. The latter becomes useful if the Legendre condition is met. Given what Moser knows at the time, in particular the issues arising in variational problems in infinite dimensions, his conclusion—"...certainly not suitable for an existence proof..."—seems well-supported by the facts. Indeed, there are no infinite-dimensional holes in infinite-dimensional spaces and therefore there cannot be any correlation between the critical points of infinite Morse index and the topology of the infinite-dimensional space. It is obvious!

Given our present context, we can paraphrase what Moser said as:

There will never be a Floer theory!

Moser studies the variational principle[43]

$$S = \int \left(\sum y_k dx_k - \lambda(H-c)\right) dt$$

which is a slight variation of the action functional. Here S depends on a real number, denoted by λ, and loops

$$z : S^1 \to \mathbb{R}^{2n} : \ z(t) = (x_1(t), y_1(t), \cdots, x_n(t), y_n(t)).$$

The fact that Moser was able to solve a local problem about Hamiltonian systems by reducing the infinite-dimensional variational problem (locally) to a finite-dimensional problem could have sparked some second thoughts—easy to say in hindsight.[44] The crux of it is that, in the context of the jungle metaphor, Moser was unknowingly standing in front of a door, albeit hidden by vegetation, that led to a wonderful garden while proclaiming:

There is nothing to see here!

While Moser was correct that, given what was known at the time, there couldn't be a general(!) critical point theory for functionals with infinite Morse index and infinite Morse coindex, due to the "infinite-dimensional hole phenomenon," it did not exclude the possibility that there was some extra structure that would "save the day" for a certain class of problems. And as it turns out, there was—even for the problem he was considering. The extra structure just had to be found.

While the non-existence of infinite-dimensional holes in infinite-dimensional spaces looked like a kiss of death for the study of smooth functions with infinite Morse index, it turned out that there was for many of the action principles a very subtle structure in the problem which had been completely overlooked. Fleshing out the details took quite some time. But in essence the structure made it possible to define the difference of the dimensions of suitable infinite-dimensional spaces of the loop space. In some cases this difference is an integer $\ldots, -2, -1, 0, 1, 2, \ldots$ and in the more notorious cases something funny happens: One can distinguish the dimensions only as cyclic numbers in some $\mathbb{Z}_{2N}$ where the even number $2N$ depends on the geometry of the phase space.

For example for the two-sphere S^2, which can be equipped with a symplectic structure ω, there is this little bit of structure needed to talk about the difference of dimensions of certain infinite-dimensional subsets of the space of loops $\mathrm{Loop}(S^2)$. To do so we need a specific mathematical construction, which utilizes the additional piece of structure. It turns out that one has to pick some auxiliary data to carry out such a construction. This leads to some specific ambiguity which is intrinsic to the problem, i.e. one can show that there is no clever idea to get rid of the difficulty. The result is that we can only talk about the difference of dimensions up to a mistake which is a multiple of 4. If several persons carry out the construction independently we might get a list describing the difference of dimensions of the same sets: $1, -3, 9, 21$. We note that the numbers always differ by a multiple of 4. Hence, mathematicians identify numbers differing by a multiple of 4 and view the dimension difference as an element in $\mathbb{Z}_4 = \{\bar{0}, \bar{1}, \bar{2}, \bar{3}\}$, the cyclic numbers mod 4. For other spaces 4 will be replaced by another even number.

3.1. Rabinowitz's Breakthrough

While the action functional used by Moser in his local studies was considered useless for global questions, a breakthrough on a much harder problem soon followed. In April 1977, Paul Rabinowitz, a professor at the University of Wisconsin at Madison, submitted a paper to the Communications on Pure and Applied Mathematics, the house journal of the Courant Institute. Wherein he showed the existence of periodic orbits for a semilinear wave equation problem, [**Ra1**]. This is a nonlinear partial differential equation, which can be viewed as the equation for a critical point of an action functional.

Rabinowitz, a former graduate student at NYU's Courant Institute, had obtained his Ph.D. in 1966 with Moser as his advisor. The main result of his thesis topic—"Periodic Solutions of Nonlinear Hyperbolic Differential Equations" [**Ra0**]—was concerned with a so-called bifurcation analysis of a nonlinear wave equation

FIGURE 3. Paul Rabinowitz.

problem. In this case, one has a small parameter ε and an equation depending on the parameter. For $\varepsilon = 0$ the solutions are known and one studies the question of which solutions can be continued to the nonlinear problem when ε varies.

In a sense, Rabinowitz's 1977 paper was a continuation of his thesis [**Ra0**]—but instead of a (local) bifurcation analysis, one looks for global periodic orbits, which should not be possible, according to Moser. Seeing Rabinowitz's work, Moser realized he was mistaken and he recognized Rabinowitz's result as a major mathematical advance. He asked Rabinowitz to have a look at finite-dimensional Hamiltonian systems, with fortuitous results—that is, the paper [**Ra2**] and soon thereafter [**Ra4**].

The breakthrough result in [**Ra2**] shows the existence of a periodic orbit for a Hamiltonian system in $\mathbb{R}^{2n}$ on a prescribed star-shaped compact energy surface. More precisely, given a smooth Hamiltonian function $H : \mathbb{R}^{2n} \to \mathbb{R}$ one would like to find, for given prescribed energy $E \in \mathbb{R}$, a solution $z : \mathbb{R} \to \mathbb{R}^{2n}$ and a real number $T > 0$ satisfying

$$\begin{aligned} &\dot{z}(t) = X_H(z(t)) \text{ and } H(z(t)) = E \text{ for } t \in \mathbb{R} \\ &z(0) = z(T). \end{aligned} \tag{14}$$

In other words, z is a T-periodic solution of the Hamiltonian system on a prescribed energy surface. While the energy E can be prescribed, it is part of the problem to find the period T. In order that solutions for such a problem exist, one needs some assumptions. The condition Rabinowitz imposes on H and E is the following:

(STAR-SHAPED). The energy surface $\Sigma_E = \{z \in \mathbb{R}^{2n} \mid H(z) = E\}$ is bounded (compact), regular, encloses the point 0 and $dH(z)(z) > 0$ for all $z \in \Sigma_E$.

While the question about the existence of periodic orbits of Hamiltonian systems were not understood at all upon the writing of [**Ra2**] (given the above generality), it is easy to see that without the compactness or regularity assumption there are counterexamples. But this was not enough and the important requirement is the star-shapedness assumption $dH(z)(z) > 0$ for $z \in \Sigma_E$.

Today, due to results by Ginzburg [**Gi1**], and Ginzburg/Gürel [**GiG1**], and independently by Herman [**He**], one knows that there are regular smooth compact

Hamiltonian energy surfaces in $\mathbb{R}^{2n}$ for $n \geq 3$ having no periodic orbits. In $\mathbb{R}^4$ there are also counterexamples for $H \in C^2(\mathbb{R}^4, \mathbb{R})$, and it is not known if there are smooth counterexamples.[45]

Formally, the problem (14) is variational. In order to find periodic orbits Rabinowitz' proof uses critical point theory for finite-dimensional approximations.[46]

One can show that the functional used by Rabinowitz does not admit a global finite-dimensional reduction! For these approximations one needs to find suitable critical points, where the Morse indices are correlated with the dimension of the approximation. But the values of the critical levels for these approximations have to be estimated carefully. One has an action of S^1 on $C^\infty(S^1, \mathbb{R}^{2n})$ by translation on the domain and the action functional to be studied is invariant under the S^1-action.

More specifically one can transform the search for a periodic orbit on a prescribed energy surface into the search for a critical point of a smooth map

$$\Phi : E \to \mathbb{R}$$

defined on a (separable) Hilbert space. These spaces are an infinite-dimensional generalization of the finite-dimensional Euclidean spaces. The map Φ has 0 as a critical point and $\Phi(0) = 0$. A priori all critical points have an infinite Morse index and Morse coindex. The problem is set-up in such a way that if one can find a critical point other than 0 the original problem—namely, finding a (nontrivial) periodic orbit on a prescribed energy surface—has been solved. Unfortunately Φ is a variation of the functionals which Moser considered useless. However, at this point Rabinowitz has already successfully studied the wave equation problem, which is even harder. He attacks the problem by constructing suitable finite-dimensional approximations. Rabinowitz's procedure, a balancing act, requires hard work in the areas of analysis and algebraic topology.[47]

Having introduced this successful idea to the study of Hamiltonian systems, Rabinowitz probes further and produces the paper [**Ra4**], which gives additional criteria for a compact regular energy surface to have nontrivial periodic orbits.

3.2. Weinstein's Conjecture

At the same time that Rabinowitz proved his breakthrough result about periodic orbits on prescribed star-shaped energy surfaces, Weinstein proved a somewhat weaker theorem in [**We4**].

FIGURE 4. Alan Weinstein.

He showed that if the compact regular energy surface Σ_E bounds a strictly convex domain, then there exists a periodic orbit on Σ_E. His approach is quite different from that of Rabinowitz, and quite tricky. The starting point was a

1948 result of Herbert Seifert—which in turn was based on a 1927 result, [**B3**], by George Birkhoff[48]—wherein Seifert studies the existence of periodic motions for certain mechanical systems, [**Se**]. It is a problem in Lagrangian dynamics and can be interpreted in some sense as a Riemannian geodesic problem. Weinstein generalizes Seifert's result to a type of Finsler geodesic problem.[49] Then the Hamiltonian problem can be reduced, given the convexity assumption, to a Finsler-type geodesic problem of twice the dimension solvable by the generalization of Seifert's result.

When Rabinowitz submitted his second paper, [**Ra4**], studying periodic orbits, to the Journal of Differential Equations, the editors picked Weinstein as referee. Weinstein did an extraordinary job not only refereeing the paper, but also in coming up with a conjecture about the conditions needed to guarantee the existence of periodic orbits on a prescribed energy surface. This is a very important conjecture and was published, [**We6**], in the same volume as Rabinowitz's result [**Ra4**]. The importance of what is now commonly referred to as the Weinstein conjecture[50] first came from the fact that periodic orbits are important in dynamics.

The first breakthrough on the Weinstein conjecture is due to Claude Viterbo, [**Vi1**], who showed that the Weinstein conjecture holds for contact type compact hypersurfaces in the standard symplectic vector space $(\mathbb{R}^{2n}, \omega)$. Hofer and Zehnder, [**HZ1**], gave a different proof of Viterbo's result highlighting a new phenomenon—the almost existence phenomenon for period orbits—which was refined by Michael Struwe in [**St1**]. Hofer later proved the Weinstein conjecture for S^3 and many other cases, [**H0**].[51] Finally Taubes proved the Weinstein conjecture for all Reeb vector fields on closed three-manifolds, [**Tau1**].

There are also results in higher dimensions. Sometimes one can use Gromov-Witten theory—related to type IIA string theory—to prove cases of the Weinstein conjecture, but otherwise the conjecture is wide open. The starting point for this connection predates Gromov-Witten theory and can be found in [**FHV**] and [**HV**]. The latter shows that the existence of a nontrivial moduli space of pseudoholomorphic spheres with suitable properties associated to a symplectic space implies the Weinstein conjecture for contact type hypersurfaces in that space. Once the Gromov-Witten theory was established, one had a language to better talk about such moduli spaces and it turned out that the Weinstein conjecture for contact-type hypersurfaces was implied by certain properties of the Gromov-Witten theory. This was made precise later on in [**LT**].

FIGURE 5. Ivar Ekeland.

Starting with the work of Ekeland and Hofer on symplectic capacities, [**EH1**, **EH2**], followed by the development of symplectic homology,[**CFH**, **FH**], and by the development of symplectic field theory (SFT), [**EGH**], it became clear that the combinatorial structure of periodic orbits can be used to define symplectic invariants and obstructions. This added to the importance of the Weinstein conjecture, since periodic orbits had turned out to be the building blocks for many of the algebraic theories capturing symplectic phenomena in a similar sense as algebraic topology captures properties of topological spaces.

3.3. Ekeland's Morse Theory

Inspired by Rabinowitz's and Weinstein's contributions, Francis Clarke, gave an amazingly short proof of the Weinstein conjecture for the convex case in [**Cl**]. It relied upon the idea that an equation of the form $L(u) = F(u)$ can be solved, provided we can solve $L^{-1}(v) = F^{-1}(v)$, and assuming that F and L can be (in some sense) inverted. Variations of this remain true even if some of the operators L and F can only be "partially inverted" and the "slack" is being picked up in form of a Lagrange multiplier. If the original problem is coming from a variational problem, it is likely to be true for the new problem. For the question at hand, the new problem can be solved by more classical methods. One needs a hypothesis to get an easier formulation, which is accessible to more classical methods, and it turns out that the convexity assumption allows just that.

From today's perspective, convexity is an odd assumption since the notion is not symplectically invariant. However, convex Hamiltonian systems have very special properties and it would be good to have a symplectically invariant formulation which captures the special features of convexity.[52]

Ekeland[53], working in nonlinear functional analysis and critical point theory, and later in mathematical economics, was similarly inspired by Rabinowitz's work on Hamiltonian systems, and Clarke's approach.

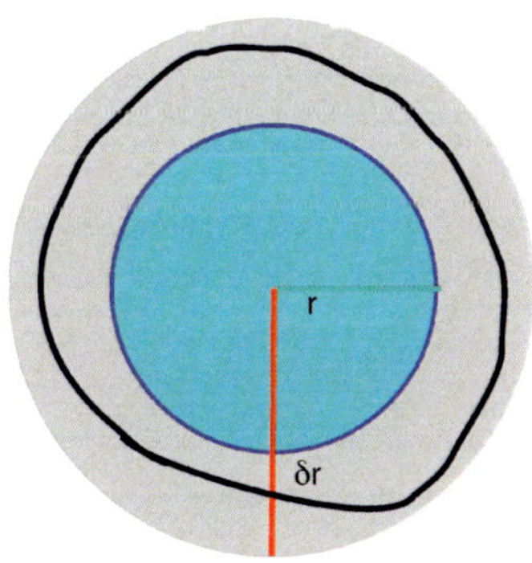

FIGURE 6. The energy surface (bold line) is pinched between a ball of radius r and a ball of radius $\delta \cdot r$.

After some joint papers with Clarke, Ekeland—in a celebrated paper with Jean-Michel Lasry—showed that on a strictly convex energy surfaces in $\mathbb{R}^{2n}$ there are at least n geometrically distinct periodic orbits, provided a certain pinching condition is met. That is, the $\sqrt{2}$-pinching condition: If the energy surface Σ in question is the boundary of a strictly convex set C in $\mathbb{R}^{2n}$, then the existence of numbers $r > 0$ and $\delta \in (1, \sqrt{2})$ such that $B^{2n}(r) \subset C \subset B^{2n}(\delta \cdot r)$ implies the existence of at least n geometrically distinct periodic orbits on Σ.[54] Here $B^{2n}(a)$ denotes the open Euclidean ball of radius a.

FIGURE 7. Yiming Long.

This paper prompted research by many others. For instance, an open question arose as to whether on a strictly convex energy surface in $\mathbb{R}^{2n}$, without any other additional hypothesis, there are always at least n periodic orbits. Investigations in this direction were initiated in [**ELa**] and [**EH3**], followed by work of Yiming Long and his school starting with the striking paper [**LZ**].

Depending on whom one asks, different conjectures about convex energy surfaces are being stated. Here is the most optimistic expectation:

CONJECTURE:. For a strictly convex energy surface in $\mathbb{R}^{2n}$ there either exist precisely n geometrically distinct periodic orbits or infinitely many.

While one of the original conjectures was that under the above assumptions there should be at least n, a result by Hofer jointly with Wysocki and Zehnder, [**HWZ1**], showed that the conjecture is correct for $n = 2$,[55] which then led to the generalization for $n > 2$. Recent progress was achieved by Erman Çineli, Viktor Ginzburg and Başak Gürel in [**CGG**]. The authors prove that in general there are at least n periodic orbits on a strictly convex energy surface in $\mathbb{R}^{2n}$. In the case that the energy surface is in addition centrally symmetric and the Reeb flow is non-degenerate there are either n or infinitely many periodic orbits.

FIGURE 8. Erman Çineli, Viktor Ginzburg and Başak Gürel.

In [**Ek1**] Ekeland uses the convexity assumption to describe the structure of the periodic orbits in terms of a more classical variational principle following Clarke's original idea. This allows him to apply Bott's generalization of Morse theory, which instead of non-degenerate critical points allows non-degenerate critical manifolds. In this case, due to an action of S^1 on the loop space, the non-degenerate submanifolds are circles. With this being set up one can introduce transversality arguments to show, first, that for a generic strictly convex energy surface all critical submanifolds are non-degenerate. Next one can show that generically the Morse inequalities imply there must be infinitely many geometrically distinct periodic orbits, addressing the intrinsic fact that given a periodic orbit also multiple covers of them appear as critical points. So we always have infinitely many critical points, but they might all be coming from the underlying prime orbit (such an issue also occurs in the Riemannian closed geodesic problem). This makes it difficult to show that there are more than one geometrically distinct orbits. Ekeland succeeded in showing that generically there are infinitely many prime periodic orbits.[56]

Another interesting story in the "convex thread" is the Viterbo conjecture, [**Vi3**]. As proved by Artstein-Avidan, Karasev and Ostrover, [**AKO**], this conjecture implies the classical 80-year-old Mahler conjecture. A version of the Viterbo conjecture estimates the action of the smallest periodic orbit on the boundary of a strictly convex subset C in terms of the volume of C. The conjecture states:

VITERBO CONJECTURE:. For a strictly convex smooth hypersurface Σ, bounding the convex set C the following equality holds

$$\mathcal{A}_{\mathbb{R}^{2n}}(P) \leq (n! \cdot \mathrm{vol}(C))^{\frac{1}{n}}$$

where "vol$''$ denotes the Euclidean volume and P is the periodic orbit on ∂C with the smallest action.

FIGURE 9. Shiri Artstein-Avidan, Yaron Ostrover and Pazit Haim-Kislev.

Even though this is a beautiful conjecture which had been very influential for some important developments in symplectic geometry, it turns out—after 24 years and many attempts of proving it—to be false. Even computer programs had been written to study this problem, but for a while seemingly all the findings seem to be supportive of the conjecture. Ultimately a candidate for a counterexample was found by a computer code. A rigorous proof was provided in a recent preprint by Pazit Haim-Kislev and Yaron Ostrover, [**HO**]. It is still possible that the Viterbo Conjecture is correct when one assumes that the convex domain is centrally symmetric. Such a version would still be sufficient to imply a famous conjecture by Kurt Mahler from 1939.

In summary, the convexity assumption allows a more classical critical point approach in these particular circumstances—one can show that convex sets have more interesting symplectic features and properties than expected. Still, the situation is unsatisfactory, since convexity is not a symplectic notion. This raises the question of whether there is a suitable symplectic property that captures the special features of convexity.

3.4. The Beginning of Global Methods

While quite a number of researchers understood the importance of Rabinowitz's contribution, the nature and extent of the understanding varied based on their field's "philosophy." The problem of finding periodic orbits was a classical one and Rabinowitz made a significant step forward. Clarke, a specialist in optimization and convex analysis, quickly realized that methods in his tool box produced a very short $2\frac{1}{2}$-page proof of the Rabinowitz result in the convex case—whereas Weinstein needed 12 pages. Although Clarke's result was weaker than Rabinowitz's, it led to the recognition that convex Hamiltonian systems have special properties, subsequently exploited by quite a number of researchers.

And one could also approach it from a different angle: Rabinowitz had demonstrated there was a novel class of abstract problems that now could be dealt with, and which would have applications to various directions in the theory of (partial) differential equations. Herbert Amann, who liked to methodically develop whatever he was thinking about, was motivated by Rabinowitz's two papers [**Ra1**, **Ra2**]. He immediately recognized the importance of the development and introduced what he called a "saddle point reduction," [**Am**]. This is a variational version of the Lyapunov-Schmidt reduction. It allowed abstract variational problems of the "untreatable kinds" to be reduced to finite-dimensional problems, provided certain technical hypotheses are met. Amann illustrated the ideas with the types of problems considered by Rabinowitz, but with different hypotheses.

FIGURE 10. Charles Conley and Eduard Zehnder (under steam).

The applications to Hamiltonian systems enticed Amann's colleague, Zehnder, and they developed the methods further, in [**AZ**]. The basic scheme is to consider time-dependent Hamiltonian systems on $\mathbb{R}^{2n}$, which have a rest point, say 0, where one knows the higher order behavior at 0, as well as the asymptotic behavior at infinity. The behavior at 0 and ∞ is quantified by some invariant which is introduced.

Then they are able to show the existence of certain nontrivial periodic orbits. In their paper the asymptotic behavior at 0 and ∞ is assumed to be time-independent, which is quite restrictive.

After the saddle point reduction, Amann and Zehnder made use of the fact that what's left is a finite-dimensional problem to which many methods can be applied. And then in 1978, Charles Conley—one of Moser's first Ph.D. students—published his celebrated work, [**C2**], now called Conley index theory. Amann and Zehnder combined their methods with Conley's ideas, resulting in the work [**AZ1**].

As Zehnder remembered in [**Ze0**]: "Conley was an extremely unconventional, sometimes eccentric man, who created his own mathematics. You never quite knew whether a statement was a joke or meant seriously. He could quote Mark Twain in every situation. He had learned Twain's work by heart while serving four and a half years in the air force, guarding a junk yard in England."

The main objects of Conley's theory are the closed invariant subsets of topological flow. The theory can be thought of as a generalization of Morse theory. The latter can be understood as the study of gradient-like flows and using Conley's theory, applied to such flows, one can draw the same conclusions. But, of course, Conley index theory can be applied to arbitrary flows, not only gradient flows. And indeed, Floer later derives the properties of the Witten complex, discussed in Section 4.3, in the finite-dimensional Morse theory case from the Conley index theory, in [**F3**].

But first comes Zehnder's collaboration with Conley, [**CZ3**], which resulted in the Conley-Zehnder index.[57] This construction allowed time-dependent asymptotic data and removed the previous restriction of the Amann-Zehnder work.

Part 2

Progress

CHAPTER 4

From Duisburg to Berkeley

Even as a youngster, Andreas was an "Überflieger"—a high flyer. But he kept to himself. "He was always thinking," said his brother Detlef, who was one year and nine months older. "Thinking, thinking, thinking." One day, Andreas' mother asked Detlef, then about four, to count some toys. Not yet enrolled in school, he was already a good counter, and his mother wanted to show off his talents for relatives. Detlef started: "One, two..." Andreas, sitting off to the side, all of a sudden said: "Five!"

"I had to count," Detlef recalled. "But here was this small boy, he did not have to count. He just saw the amount." Detlef continued, "It took me quite a while to get over it, that is why this experience was seared into my memory. How can a 3-year-old simply know the right answer, without counting? This shows a flash of what was there. He was wired differently. It wasn't learned. It was inside him from the beginning."

FIGURE 1. Andreas, about age two.

At the end of World War II, Andreas' parents had lived in East Berlin, in Treptow, part of the Soviet sector. His father, Bruno, was very conservative. An electrical engineer from a Catholic family, he was forced to relocate and work in Russia—he was forcibly deported; German expertise in science and engineering were urgently needed. Their mother, Marlies, from a liberal Protestant background, was a housewife despite having trained as a chemist. She was given 24 hours to decide whether to join him; she did. In Russia, Marlies had two children, a daughter who died shortly after childbirth, followed by their eldest son, Rainer, who was born in Gorki in 1948. The family returned to Germany in the early 1950s. Detlef was born in 1954, and Andreas on Thursday, August 23, 1956, in Duisburg. Their father died of a stroke in 1967 at age 57. Marlies, who suffered from depression, was at once a strong woman and raised the three boys alone.

Andreas was a straight-A student throughout high school at the Gymnasium; all that thinking had launched him on a heady path. Next he studied electrical engineering at RWTH Aachen University. In his first year he switched to physics,

FIGURE 2. Andreas with his mother and brother Detlef in Zandvoort, the Netherlands, where they vacationed every summer.

and then moved to the Ruhr-Universität Bochum where he pursued mathematics. After receiving a diploma thesis[58] topic from Professor Ralph Stöcker—in algebraic topology, probing Poincaré duality spaces[59]—Floer wasn't seen for quite a while. When he reappeared to hand over the manuscript, he reportedly declared: "This is all that one can say about the topic." Floer's verdict was that the subject was too esoteric. Nevertheless, his encounter with algebraic topology proved important for later developments.

Having finished his diploma thesis in 1982, Floer pursued graduate studies. He wanted to see the world. Rolf Kaiser, a university chum of Detlef's, recalled that Andreas was keen to create some distance with the "Mutterhaus" (the motherhouse). And he wanted to do contemporary mathematics research. He started graduate school at the University of California, Berkeley, in fall 1982. That Andreas was accepted at Berkeley would not have been surprising since he was an exceptionally strong student (and he had a 10-month stipend from Germany).

* * *

Floer followed his curiosity—incorporating new ideas, accumulating a body of knowledge—and positioned himself to take advantage of the evolving landscape. He worked at the intersection of several fields, and over time introduced what today are called Floer-theoretic methods.

At Berkeley, Floer landed in clover. His arrival coincided with the opening of the Mathematical Sciences Research Institute (MSRI), with Shiing-Shen Chern as its founding director. Mathematicians convened downtown at the institute's initial home on Fulton Street. "There was a lot of energy," said Dan Freed, who was then

FIGURE 3. S.S. Chern.

also a graduate student—every class, every seminar, had an audience of at least 50. Later, MSRI moved to the top of a hill above Berkeley's main campus. Upon an endowment gift in 2022, it became the Simons Laufer Mathematical Sciences Institute.

4.1. The Donaldson Seminar

The MSRI inaugural program, from August 1982 through July 1983, addressed nonlinear partial differential equations.

FIGURE 4. Shing-Tung Yau and Michael Atiyah.

It counted among its esteemed attendees a number of future Fields Medal and Abel Prize laureates. Organized by Alexandre Chorin, Isadore Singer (of Atiyah-Singer fame), and the freshly minted Fields medalist Shing-Tung Yau, the program focused in part on the big news of the day: a recent preprint by Simon Donaldson (a Ph.D. student of Atiyah and Floer's junior by about twelve months) [**Do**]. Donaldson's main result—deemed "one of the most surprising results in topology" [**Ki**]—earned him a Fields Medal four years later.

FIGURE 5. Simon Donaldson.

Karen Uhlenbeck and Michael Freedman ran a seminar on Donaldson's theorem during the fall term—Dan Freed took notes (which ultimately resulted in a book, co-authored with Uhlenbeck [**FU**]). The theorem shows that closed topological four-manifolds admitting smooth structures must have certain algebraic topological properties. And, using work by Freedman, [**Fr**], who also won a Fields Medal in 1986, the result implies that there are closed topological four-manifolds which do not admit smooth structures.

FIGURE 6. Karen Uhlenbeck, Michael Freedman, Daniel Freed.

Looking back, Uhlenbeck doesn't recall much about the seminar. She does remember Floer. "He was a real intense guy, that was clear," she said. "Bright and ambitious. Not easy going, but he didn't stand out that way. Some mathematicians are friendly and sociable, some are awkward and not comfortable in their skins."

She continued, "Andreas made a big impression on me, because I do remember that he gave a presentation. And I remember that, in fact, his presentation was on the equivariant connections—it was a question of the topology that comes from the equivariant connections. And I do remember that it was wrong. These equivariant theorems are very tricky in topology. I don't remember why it was wrong. I just remember that we had to redo it, Dan and I had to fix up his proof."

FIGURE 7. Andreas, 1982—a visa application photo.

Floer wasn't on top of everything; he was gradually wandering into the jungle. Over the course of that term, he learned about Uhlenbeck's bubbling-off analysis, Taubes' gluing analysis, and more generally about the Yang-Mills equations and gauge theory—all ideas that later became important to his work.[60] He was clearly in his element in Berkeley. However, he was only meant to be there for ten months. He was expected to return to Germany in July 1983 in order to continue his alternative military service at the Technische Hilfsdienst. By early January, he wrote a letter requesting an extension:

"There is the opportunity through employment at the university in Berkeley to continue my studies up to a Ph.D. Since the University of California is one of the best places in my area of research in the world this would be very important for my professional future."[61]

His wish was granted for fall 1983 and spring 1984. Around the same time he wrote to Ralph Stöcker with whom he had written his diploma thesis.[62] He mentioned that Cliff Taubes would be joining the Berkeley faculty in the fall of 1983, and that he found his work very interesting (off to the side, he complained that "Always a third of the Berkeley faculty is on leave"). He also mentioned that he was preparing for the Ph.D. qualifying exam: "I decided to go through the formalities of the Ph.D. program." he wrote. "Before I return to Germany I want at least reach the level of Ph.D. candidate which might hold some advantages for later." (See the Appendix on page B for Floer's qualifying exam topic list.)

4.2. The Browder Conference

In July 1983, the thirty-first Summer Research Institute of the American Mathematical Society convened at Berkeley. The topic was "Nonlinear Functional Analysis and Its Applications," organized by a scientific committee chaired by Felix Browder (other committee members were Haim Brezis, Tosio Kato, Jacques-Louis Lions, Louis Nirenberg and Paul Rabinowitz).

More than 200 participants attended, from some 20 countries. The program was packed and compelling; 13 series of expository lectures and 115 one-hour research lectures. Several lectures chronicled recent breakthroughs that would become pivotal for Floer's research[63]: Rabinowitz on "Minimax methods for indefinite functionals"; Zehnder on his work with Conley, "A global fixed point theorem for symplectic maps and subharmonic solutions of Hamiltonian equations on tori"; Ekeland on, "An index theory for periodic solutions of convex Hamiltonian systems"; and Weinstein, "On extending the Conley-Zehnder fixed point theorem to other manifolds."

These were exciting times. Interesting mathematics was in the air, with an almost palpable cross-pollination of ideas. It was at the Browder conference that Zehnder introduced Floer to Helmut Hofer. Hofer then introduced Floer to Abbas Bahri, already an accomplished nonlinear analyst who was thinking about the Weinstein conjecture.[64] The group of them convened for several dinners, deconstructing and discussing the talks of the day. For the most part, Floer sat quietly and listened.

4.3. The Witten Effect—On Morse Theory

In August 1979, at Cargèse, Raoul Bott delivered a series of lectures on Morse theory to a group of physicists, including Edward Witten. The lectures addressed equivariant Morse theory and its pertinence to the Yang-Mills theory on Riemann surfaces. Bott recalled:

"I was reporting on joint work with Atiyah to a group of very bright physicists, young and old, most of whom took a rather detached view of the lectures. 'Beautiful, but oh so far from physics' was [future Nobel laureate] Kenneth G. Wilson's reaction... On the other hand, Witten followed the lectures like a hawk, asked questions, and was clearly very interested. I therefore thought I had done a good job indoctrinating him in the rudiments of the half-space approach, etc. so that I was rather nonplussed to receive a letter from him eight months later, starting with the comment, 'Now I finally understand Morse theory!'."

Bott came to consider Witten's contribution as one of the four crucial developments in Morse theory. [65] He described, [**Bo2**], the progression as:

Morse → Thom → Smale → Witten → Floer[66]

But this considerably oversimplifies the landscape leading to Floer theory, as we shall see.

FIGURE 8. Edward Witten.

Witten's paper on "Supersymmetry and Morse Theory" [**Wi**] appeared in 1982. It was much talked about, so Floer might have first set eyes upon it soon after he arrived in Berkeley.[67] The paper spotlighted certain facts known to the likes of John Milnor, Stephen Smale and René Thom, but seemingly they did not consider it too important. These underappreciated ideas, however, provided a way to define so-called homology groups, which are an important construction in algebraic topology, using tools from Morse theory/dynamical systems rather than using the typical methods. Since the typical methods of algebraic topology seemed more efficient in the usual finite-dimensional context, the alternative way had no impact until Witten came along.[68]

Witten's viewpoint was that of a physicist. As Bott observed: "... it taught us to relate the quantum mechanical concept of 'tunneling' to topology." The importance of this Wittenian development was that the Morse-theoretic construction of homology could then be generalized to infinite dimensions—that is, to Floer theory (Whereas the algebraic topology methods could not; or it is not known how, at least.[69]).

Izu Vaisman, [**Vai**], reviewed Witten's paper in the Mathematical Reviews. In an elaborate abstract, he began by saying:

"The Hilbert space of a quantum field theory has a decomposition $\mathcal{H} = \mathcal{H}^+ \oplus \mathcal{H}^-$ into bosonic and fermionic states, and a super-symmetry is defined by Hermitian operators Q_i between the two terms of this decomposition, which satisfy various conditions, including commutation with the Hamiltonian operator H. In this context, it is important to discuss solutions of the equation $Q_i|\Omega\rangle = 0$, as well as their nonexistence. This can be seen as an index problem for operators on a differentiable manifold. By this approach, supersymmetry is related to differential topology, and this ample paper discusses various such relationships."

He continued—with the salient point for our purposes—to say: "Namely, the simplest supersymmetry algebra yields a new insight into Morse inequalities and generalizations, and a more elaborate algebra is related to fixed point theorems for Killing vector fields. Operators which are usual in differential geometry, and their deformations, are involved."

Witten's paper was received with interest by many mathematicians, but not all—it was seen as a statement that these ideas were important in physics and therefore only mathematicians of a certain bent took note. The landscape changed due to greater intermingling between mathematicians and physicists, due in great part to Michael Atiyah. That Witten's paper rather quickly attracted the likes of Bismut, [**Bis**], Helffer and Sjöstrand, [**HeS**] added substantially to the popularity of Witten's ideas among certain groups of mathematicians.[70]

* * *

As we recall from Chapter 1, Morse theory establishes, under certain conditions, for a given smooth function f defined on a smooth space Λ

$$f : \Lambda \to \mathbb{R}$$

a relationship between the critical points of f and the topological properties of the space Λ. This is algebraically captured by the Morse equality which we have seen several times before

$$P_f(t) = P_\Lambda(t) + (1 + t) \cdot Q_f(t).$$

This formula already boils Morse theory down to an algebraic relationship and in some sense can be viewed as a count of critical points in terms of the global properties of the topological space Λ measured by certain tools from algebraic topology.[71]

It is worth contemplating what Floer took away from Witten's paper, [**Wi**]. Were these ideas triggers towards the development of Floer theory, or was Floer already well on his way? Hofer initially believed that Floer knew of Witten's work when he had the initial idea for Floer theory—this seemed to be the natural order of things. However, in reading Floer's papers again, especially the references, some doubts arose. It is clear that, at the latest, Floer was aware of Witten's paper in the first half of 1986, and he observes the relationship to Conley's index theory due to his Ph.D. work on the Arnold conjecture for surfaces.

Conley's theory is a homotopy theoretic theory of flows, which allows one to use the tools of algebraic topology ([**C2**] see also [**Sa1**, **Sa2**]). Morse theory for a function f can be understood in terms of Conley's theory applied to the so-called gradient flow associated to f around Conley's idea of a Morse decomposition for topological flows.[72] The idea of a Morse decomposition for topological flows, applied to gradient flows, also implies the same results—important for the development of Floer theory—which can be deduced from Witten's paper.

Later, after having developed his seminal ideas, Floer writes down this observation in [**F3**]. He also remarks in [**F2**]: "With the field $\mathbb{Z}_2$ replaced by $\mathbb{R}$, it has been studied by Witten in his recent paper [27], which partly prompted our approach to the Morse theory of the symplectic action." What Floer means by "... partly" is not entirely clear. The impact of Witten's approach on Floer will be discussed in further detail in Chapter 9 and in particular in Section 9.3. But before Floer even had started his work, Conley's index theory for topological flows will play an important role in the Conley-Zehnder breakthrough result introduced in Section 6.4.

As we shall see, Floer needs another trigger. It will come in the form of Gromov's pseudoholomorphic curve theory against the backdrop of Zehnder's philosophical elaborations.

CHAPTER 5

The Magic

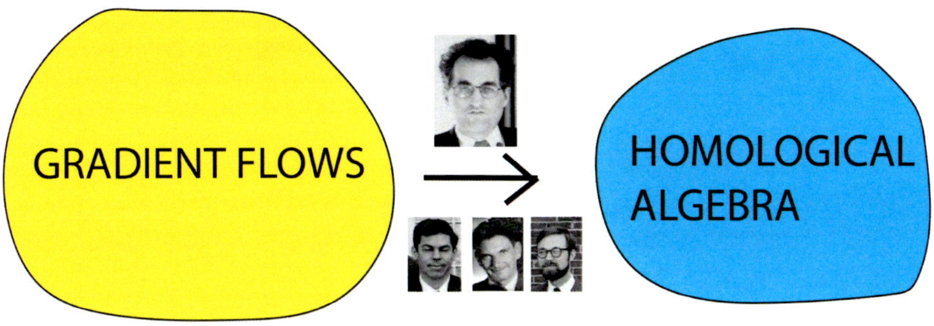

FIGURE 1. The plan—Witten, Smale, Thom, Milnor.

In the 1950s, the field of homological algebra underwent a revolutionary change. Among the revolutionists at the center was the famous mathematician Henri Cartan. Some thirty years later, in 1980 when Cartan was awarded a Doctor Honoris Causa from Oxford, he still was in awe about the field he had transformed. As he himself said[73] regarding the main structural equation in homological algebra: "..if I could only understand the beautiful consequences following from the concise proposition $d^2 = 0$."

The gist of the mathematical object that Cartan was marveling about—chain complexes—can be conveyed via the interplay of two basic ideas: communication channels and superposition.

We shall use many pictures to make this powerful idea digestible to the reader. At the heart of the endeavor is the goal to organize mathematical data. The idea was born in the field of topology, but was so successful that it also became crucial in other mathematical disciplines and even in applications, for example the theory of error correcting codes—classical and quantum. See for example [**HaH**] for the quantum theory.

Of course, this was helped by the fact that mathematical data coming from different fields has, surprisingly, certain common features, which were found when one started looking for them—a journey starting in topology.

The overall idea one should have in mind is as follows. Suppose we have a layered body of information, where we denote the different layers by $\mathfrak{I}_0, \mathfrak{I}_1, \ldots, \mathfrak{I}_n$. Without defining what we mean by information, the reader might think of $\mathfrak{I}_k$ not only as a depository of code or strings of numbers, but also as entity being able to process data. It might even have its own language for processing and communicating data.

FIGURE 2. The "stealthy information" in the square representing the box $\mathfrak{J}_k$ is enclosed by the bold black border. It contains the "image data" (thin black border) coming from the the box $\mathfrak{J}_{k+1}$.

These depositories communicate, but in a one-sided way. Namely $\mathfrak{J}_{k+1}$ communicates some information to $\mathfrak{J}_k$. Thus, we have a hierarchy, pictureed as follows.

$$\boxed{\mathfrak{J}_n} \Longrightarrow \boxed{\mathfrak{J}_{n-1}} \Longrightarrow \dots \Longrightarrow \boxed{\mathfrak{J}_0}. \tag{15}$$

We should view the whole system as a way to stratify a certain type of data and to exhibit its relationships. Moreover, it turns out that we will be able to talk, after some processing, about data proprietary to the individual "boxes" $\mathfrak{J}_k$.

The information in each box is being prepared and sent to the next box. The arrows are communication channels and we see that the box $\mathfrak{J}_k$ can only send information to $\mathfrak{J}_{k-1}$, but $\mathfrak{J}_{k-1}$ cannot send information to $\mathfrak{J}_k$.

Next we have a look at a portion of the directed information system.

$$\underset{\text{orange}}{\boxed{\mathfrak{J}_{k+1}}} \Longrightarrow \underset{\text{red}}{\boxed{\mathfrak{J}_k}} \Longrightarrow \underset{\text{blue}}{\boxed{\mathfrak{J}_{k-1}}}. \tag{16}$$

We focus on the red box and ask two basic questions:

(1) What is the totality of information received by the red box through the orange arrow? We shall call this "image data."
(2) What is the totality of information in the red box which is stealthy for the blue box, i.e. it does not make it through the blue arrow? This is the "stealthy data."

If we think more about this structure we can ask a "higher order question"

(3) What is the information in the orange box which will arrive in the blue box after being sent to the red box where it is processed and sent further? We would call information in the orange box which makes it to the blue box "pass-through information."

This is the point where we need to make a philosophical decision regarding the data or information type we would like to study and organize. Our goal is to organize data and we would like to keep it as simple as possible. For this we assume that the layereddata allows some directed communication, but the layering and the overall system is organized in such a way that (3) does not happen—no "pass-through information."

Due to the non-existence of "pass-through information" the image data arriving at the red box is stealthy, see Figure 2. The blue data in the red box—the "stealthy data"—is the one which cannot be seen by the blue box. It contains some contamination, namely the orange image data. The "proprietary data" associated to the red box is given as

$$\text{propdata}_k = \text{bluedata}_k \ominus \text{orangedata}_k. \tag{17}$$

Here the $\ominus$-sign has to be understood as some procedure special to the red box, which annihilates the contamination by producing some new data representing the "non-contaminated data." If we can come up with such a scheme we obtain for every $k \in \{0, \dots, n\}$ proprietary data: propdata_k. As it turns out homological algebra precisely does this in many mathematical contexts. The above scheme applied in a mathematical area always looks like

$$\text{mathDATA} \overset{processing}{\longrightarrow} \text{layereddata} \longrightarrow \text{propdata}.$$

For the moment a reader should view a chain complex as an abstract mathematical concept. Precise knowledge is not needed at this moment—one just should keep the previous discussion in mind.

5.1. Short Overview on Witten's Construction

The goal of Witten's construction, which we shall describe first in a heuristic form, follows the previously outlined scheme. One would like to quantify, using an algebraic language, the topological properties of a (compact) smooth space Λ.

The "processing" then consists of first picking—possibly an arbitrary choice—a Morse function $f : \Lambda \to \mathbb{R}$, and a generic metric g (a fact we shall suppress) followed by a natural[74] recipe to construct an algebraic object $(C(f), \partial)$, called a chain complex, which reflects the properties of f. As we have seen in the beginning chapters—a consequence of the birth-death mechanism—there are Morse functions which have extremely high numbers of critical points. In such cases $(C(f), \partial)$ can be quite complicated.

The chain complex encodes in an algebraic way important structural features of the Morse function f on Λ. Connecting with our previous discussion, Λ is the "mathDATA" and $(C(f), \partial)$ will be the "layereddata."[75] Note the interesting fact that there is the choice of (f, g) involved. It will, however turn out, that the end result, i.e. "propdata," does not depend on this choice. This is the reason why this overall idea is so powerful![76]

More precisely, homological algebra provides a basic algebraic procedure (some part of the magic) which can be applied to any chain complex called "taking its homology." Applied to $(C(f), \partial)$ we obtain the homology $H(f)$ associated to f and it turns out that $H(f)$ does not(!) depend on which Morse function one takes on Λ.[77] Here $H(f)$ is the "propdata." It is in fact a very a surprising fact that given a smooth space Λ, any Morse function $f : \Lambda \to \mathbb{R}$ produces the same "propdata." The "propdata" can therefore be used as an invariant describing properties of the space Λ.

Of course, it was not clear that anything like this could be done, but the possibility was buried somewhat within Milnor's, Smale's and Thom's work, and was made obvious by Witten approaching the question from a new angle. Witten derives finite-dimensional Morse theory in a different way, but it is equivalent to the usual methods.

The importance for the Floer story is that Witten's version can be generalized to the situation where we do not have flows, but rather partial differential equations with asymptotic boundary conditions. Floer zeroed in on one aspect of Witten's result: the fact that " ... the simplest supersymmetry algebra yields a new insight into Morse inequalities... "—as described in the review of Witten's paper, [**Vai**]. Floer transported Witten's idea via an analytical tour de force to infinite dimensions by breaking the symmetry of Gromov's 1985 invention of "Pseudoholomorphic curve theory," [**G**], see the Prologue on page xvi.

In order to obtain Witten's representation of Morse theory it is important to first notice properties of generic gradient flows. Data extracted from generic gradient flows is used to define what is called a chain complex which captures properties of f that have some special features due to the existence of the birth-death mechanism.

5.2. From Gradient Flows to Graphs

Previously, in Section 2.1, we already described the idea of a flow. A particular class of flows is the class of gradient flows. Starting with a smooth function $f : \Lambda \to \mathbb{R}$ we can take the derivative $df(x)$ at every point $x \in \Lambda$. Fixing as auxiliary data a Riemannian metric g we can turn the assignment $x \to df(x)$ into a vector field denoted by $\nabla_g f$. The Cauchy problem

$$\dot{x} = -\nabla_g f(x) \quad \text{and} \quad x(0) = x_0$$

defines a flow on Λ called the gradient flow—or more accurately the "minus gradient flow." It has the important property that for every initial data x the function $t \to f(t\star x)$ is strictly decreasing unless x is a rest point. In our discussion of Morse theory the water flow pictures from Chapter 1 exhibit gradient lines. Moreover, the rest points are precisely the critical points. Our procedure associates to a smooth map a dynamical system the gradient flow.

5.2.1. Rigid Flow lines.

Of interest are the bounded flow lines which are solutions $t \to t \star x$ of the gradient flow equation which converge in forward and backward time to a critical point

$$x^- = \lim_{t\to-\infty} t \star x \quad \text{and} \quad x^+ = \lim_{t\to\infty} t \star x.$$

Among these flow lines are the rigid ones, which are precisely those for which the Morse indices of x^- and x^+ are related by

$$m(x^-) = 1 + m(x^+).$$

The following figure shows a rigid flow line. Rigidity can also be understood as follows. A flow line is rigid precisely if given any point y near the flow line, but not lying on the flow line, the associated flow line $t \to t\star y$ cannot converge to both x^+ and x^-, as $t \to \pm\infty$. One can show that a gradient line $\mathbb{R} \star x$, connecting x^- to x^+, with $m(x^-) - m(x^+) = 1$ is rigid, see Figure 3.

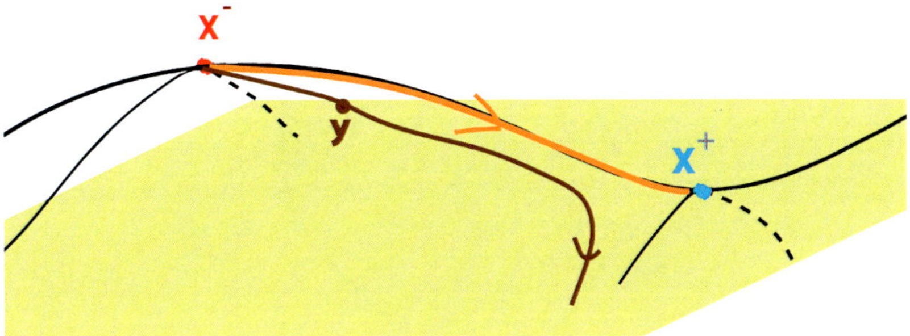

FIGURE 3. A rigid flow line $\mathbb{R} \star x$ between x^+ and x^-. The point y is close to the flow line $\mathbb{R} \star x$ and has $y^- = x^-$ but $y^+ \neq x^+$.

An interesting fact about gradient flows is that they have nice generic properties, provided that f has well-behaved critical points and the auxiliary data g, the Riemannian metric, is picked carefully[78] (see the paper by Smale,[**Sm1**]). We shall refer to a pair (f, g) where f is a Morse function and g is 'fine-tuned' as a generic pair. Rather than defining what generic means we shall list the implied properties. The main construction will capitalize on this fact. Here is a list of assumed properties on a generic pair (f, g):

(18)

(1) Every gradient line $t \to t \star x$ which is not constant, converges in forward and backward time to constant states, which are the critical points of f. Introduce

$$x^{\pm} = \lim_{s \to \pm\infty} s \star x$$

and denote by $m(x^{\pm})$ the Morse indices, viewing the rest points as critical points of f.

(2) For a non-constant state x it holds that $m(x^+) < m(x^-)$.

(3) The gradient flow is Morse-Smale and satisfies some compactness property—similar to the property (PS)—which we do not specify further.

The third property implies the second and is very important. Since it is somewhat technical to state, we omit a precise description (see [**Sm1**] for more details concerning the Morse-Smale property).

The key point we want to make is that under the assumptions (1)–(3) the following is true:

(Fact A) Given rest points $x^+ \neq x^-$ with the property that $m(x^+) = m(x^-) - 1$, the flow lines connecting x^- to x^+ are isolated and of only a finite number.

The picture the reader should have in mind is the following: Given a Morse function $f : \Lambda \to \mathbb{R}$ (on a closed manifold) we look at its finitely many critical points and the rigid gradient lines between them—recall the example given by Figure 14.

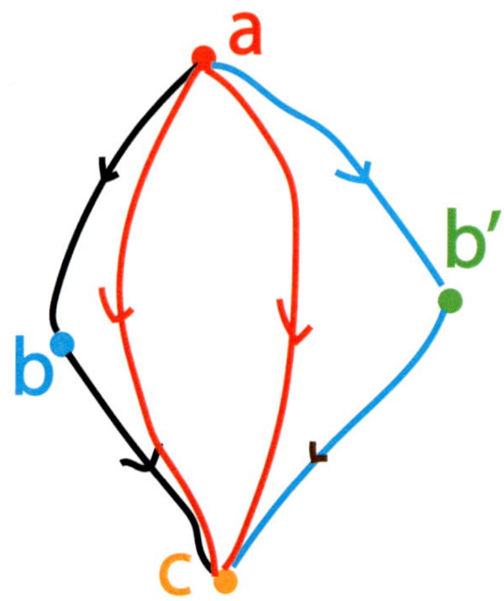

FIGURE 4. Two broken trajectories. Here we assume that the Morse indices satisfy $m(a) = m(b) + 1 = m(b') + 1 = m(c) + 2$. On the left the connection $a \to b \to c$ and on the right $a \to b' \to c$, which we shall refer to in the future as rigid broken flow line. These pairs are rigid and connected by a one-dimensional family of flow lines going directly from a to c.

5.2.2. Rigid Broken Flow Lines. As we shall see, the system of rigid flow lines has an interesting and unexpected feature and this is crucial for the Witten procedure. To explain this consider rigid broken flow lines as shown in following Figure 4. On the left we have a broken flow line going from a to b and then one from b to c. Similarly on the right in short notation (a, b', c). We also have a one-dimensional family of flow lines from a to c which, on the right and the left, approximate the broken flow lines.

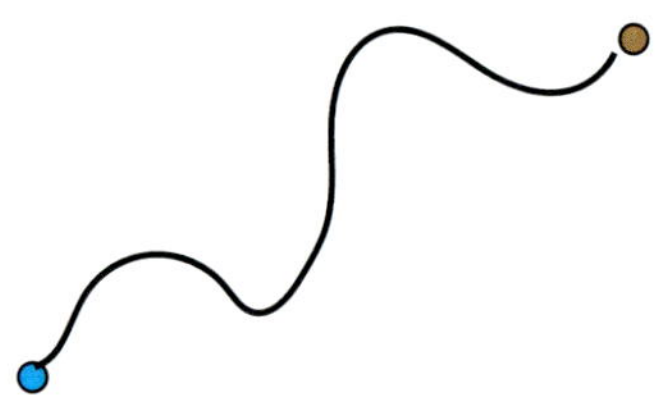

FIGURE 5. One can view the rigid, once-broken flow lines as the boundary points of a compact one-dimensional space homeomorphic to the closed interval $[0, 1]$. Since $[0, 1]$ has precisely two ends, these broken flow lines from a to c occur in pairs.

The important feature is the following: Assuming the properties listed in 18, it is true[79] that given a rigid broken flow line as shown in Figure 4, there has to be another broken flow line connected to the first by a one-dimensional family; this is captured by Figure 5.

Moreover, near a rigid broken trajectory there is precisely one of these one-dimensional families, see Figure 6 which shows two other possible configurations.

Namely, two rigid broken trajectories (A, B) and (C, D) are unless $(A = C$ & $B = D)$; thus, it is possible that we have (A, B) and (A, C), or (A, C) and (B, C).

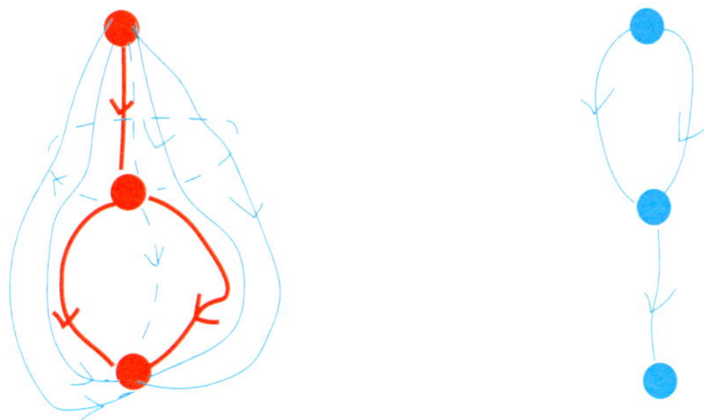

FIGURE 6. Two additional configurations of broken trajectories. On the left, the one-dimensional family is indicated and winds around the back.

Hence we have established another important fact for the generic cases we are considering:

(Fact B) Broken rigid flow lines appear in pairs. In particular the number of broken flow lines going from x to z is even!

We shall illustrate our findings by examples—associating to a Morse function the graph arising from rigid flow lines.

5.2.3. Examples: Directed Graphs from Rigid Flow Lines. Using previously discussed Morse functions as examples, we shall describe the graphs arising from rigid flow lines assuming Fact A and Fact B which hold in the generic case, see page 58 and page 59, respectively.

We have seen that we can associate to a Morse function $f : \Lambda \to \mathbb{R}$ its gradient flow (after fixing some auxiliary data). The gradient flow satisfies the assumed properties listed in 18. What sticks out from the flow lines are the rigid ones, and the fact that broken ones occur in pairs as discussed in Subsection 5.2.2. Also every such flow line starts at a critical point of some Morse index m and ends at a critical point with Morse index $m - 1$. These rigid flow lines form a directed graph as demonstrated by the following figures.

The first figure is the previously considered example of a landscape over the circle $\mathbb{S}^1$. In this one-dimensional case all eight flow lines are rigid as shown in Figure 7. One can also enhance the graph on the right by decorating vertices by the Morse indices of the critical points they represent. That would mean decorating A, B, C, D with a 1 and W, X, Y, Z with a 0.

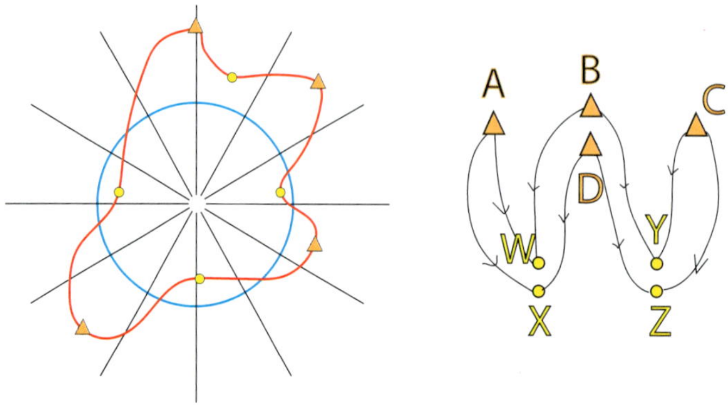

FIGURE 7. The schematic encoding of the critical points and rigid flow lines (from Figure 6) as a directed graph. This directed graph will be used later for an algebraic construction, which can be viewed as a simplified representation of $f : \Lambda \to \mathbb{R}$.

Next we consider the torus example given in Figure 8. In this case the torus is somewhat tilted and lies in $\mathbb{R}^3$ and the functional is the height function. We have four critical points where A is a global maximum ($m = 2$), B and C are saddle points ($m = 1$) and D is a global minimum ($m = 0$). The graph is already somewhat more complicated.

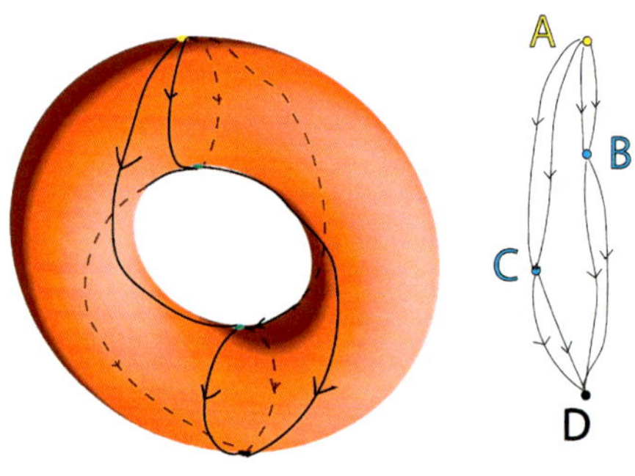

FIGURE 8. The torus in $\mathbb{R}^3$ equipped with the height function and the resulting graph.

Taking into account that the vertices represent critical points which have Morse indices we obtain decorated graphs, where the vertices have additional data attached to them—the Morse index. This is demonstrated by Figure 9.

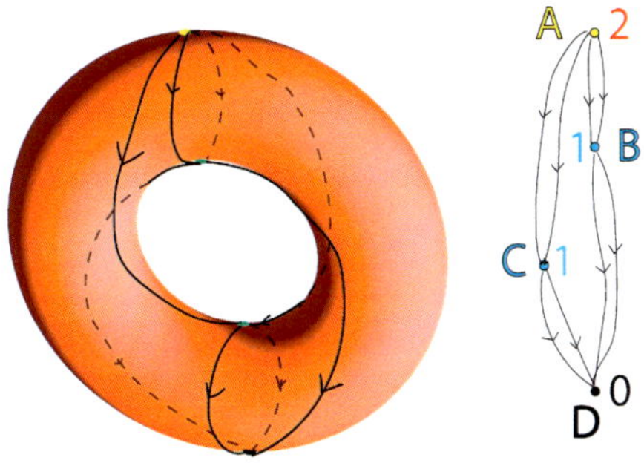

FIGURE 9. The torus in $\mathbb{R}^3$ equipped with the height function and the resulting decorated graph.

Whenever we start with a Morse function on a closed manifold Λ we will obtain a decorated finite graph—finitely many edges with finitely many directed vertices. These graphs have the special property that the pairs of broken edges which represent rigid broken flow lines occur in pairs. The following Figure shows such a graph on the left and a more schematic version on the right.

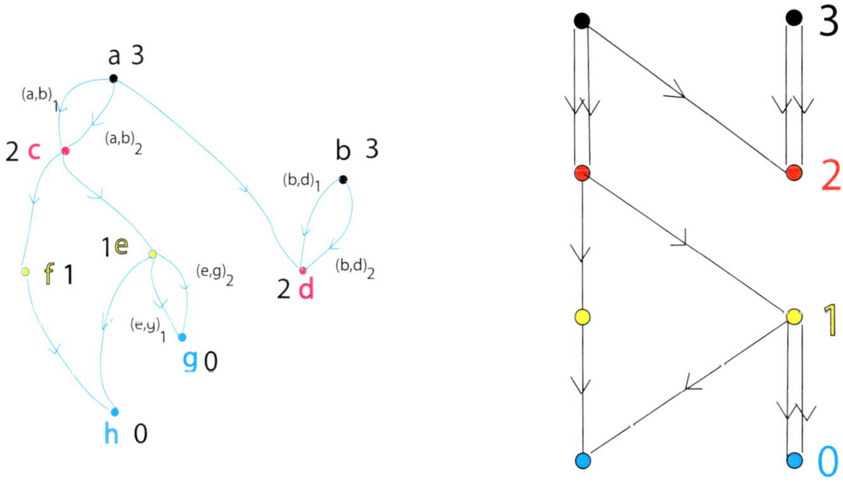

FIGURE 10. A picture of rigid gradient lines on the left and a more schematic one on the right.

In summary, given a smooth Morse function $f : \Lambda \to \mathbb{R}$ on a closed smooth space, one can construct a decorated directed graph, say G_f. Whereas in the finite-dimensional case G_f is obtained from the study of the gradient flow associated to f, Floer will have to find a way to associate to a function on an infinite-dimensional space a partial differential equation which then is needed to produce a graph of a similar kind, even if it can have infinitely many vertices and edges. This is highly nontrivial.

One of the questions which then arises is whether we can associate to G_f some algebraic object, which is independent of f when f varies over all (admissible[80])

Morse functions on Λ, and only depends on Λ. The answer turns out to be "Yes" which is some kind of a miracle, and we shall discuss this in the next section. Then one can apply essentially the same algebraic machinery which will produce an algebraic object only depending on the underlying space.

5.3. Communication Channels and Superposition

Figure 10 gives an example of the structure of the set of critical points and its rigid flow lines coming from some kind of abstract situation. It shares one crucial property with the directed decorated graphs coming from Morse theory: The number of broken edges from any vertex of index k to another vertex of index $k-2$ is even. As we have previously seen, this property is a consequence of the birth-death mechanism.

Our next goal is to turn such pictures into algebraic objects with desirable properties, which we already alluded to in a previous discussion. What we are going to do is to explain an important basic idea in homological algebra.

We begin with a quote by Charles Weibel, [**Weib**], who describes some of the history of the subject.

"Homological algebra had its origins in the 19th century, via the work of Riemann (1857) and Betti (1871) on "homology numbers," and the rigorous development of the notion of homology numbers by Poincaré in 1895. A 1925 observation of Emmy Noether [N25] shifted the attention to the "homology groups" of a space, and algebraic techniques were developed for computational purposes in the 1930's. Yet homology remained a part of the realm of topology until about 1945."

Initially homological algebra belonged to the realm of topology and it turned out to be such a successful theory that it entered other more algebraic mathematical disciplines. In some sense homological algebra was capturing a rather universal mathematical phenomenon which seemed to be present in one form or the other everywhere.

According to Weibel, Cartan and Samuel Eilenberg started (with their book [**CE**]) "...a true revolution in mathematics, and as such it was also a new beginning."

With this revolution homological algebra entered many fields of mathematics. Mathematicians who were at the center of "revolutionary changes" in the development of homological algebra, like Henri Cartan, still wondered about the power of the basic idea thirty years after the revolution.

5.3.1. Analyzing One Example. We have to address the question how to construct an algebraic object which we can assign to a directed decorated graph, e.g. to the one occurring in Figure 10.

Figure 11 should remind the reader of

$$\boxed{\mathfrak{J}_3} \Rightarrow \boxed{\mathfrak{J}_2} \Rightarrow \boxed{\mathfrak{J}_1} \Rightarrow \boxed{\mathfrak{J}_0}.$$

However there is some fine structure coming from the edges of the graph, which connect level k to level $k-1$, explaining the arrows $\boxed{\mathfrak{J}_k} \Rightarrow \boxed{\mathfrak{J}_{k-1}}$. We shall use the directed edges to construct the arrows "$\Rightarrow$."

Here is a crapshoot: We can think of the decorated vertices with index k to live on the k-th floor and the directed edges can be viewed as directed communication channels which can send signals from the k-th to the $(k-1)$-th floor. Since we already know from Morse theory that a certain number of critical points are required

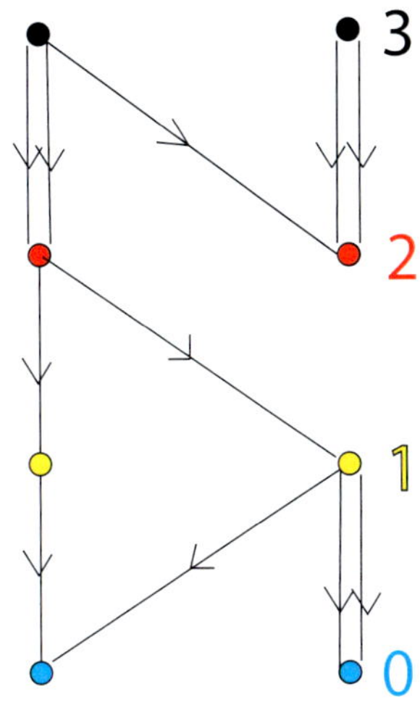

FIGURE 11. A directed graph.

for topological reasons, we could try to figure out the role played by specific critical points as follows.

In a first step assume that on the 3rd floor the two critical points are lamps. A lamp can be lit or not and the possible status of our two-lamp system can be described by the following:

FIGURE 12. The four states of a two-lamp system.

We can describe the status of a single lamp by the two numbers 0 and 1 and this allows us to represent the lighting scheme as pairs of numbers:

$$(0,0) \quad (0,1) \quad (1,0) \quad (1,1)$$

On the third floor we see four possible lighting scenarios. Now we use the cables to submit this information to the 2nd floor—alternatively we could also think that the second floor is spying on the third.

We need some rules describing the precise role of the cables.

If we consider on the third floor a light we send through all the cables starting from there a 0 if the lamp is not lit and otherwise a 1.

What do we do with the signals arriving on the 2nd floor? At any given lamp on the second floor several cables arrive and we need to consolidate the incoming data. We use the idea of superposition. We add up all signals which can be a 0 or a 1, but as addition we use the following arithmetic which incidentally is the addition a computer carries out at its basic level:

$$0+0=0, \; 0+1=1+0=1, \; 1+1=0 \tag{19}$$

This addition on the set $A = \{0,1\}$ defines what is called an Abelian group structure on A and one often denotes this group by $\mathbb{Z}_2$. More generally, a set A equipped with a binary operation "$+$" having the following formal properties—similar to the addition of real numbers or integers—is called an Abelian group:

(1) $(a+b)+c = a+(b+c)$ for all $a, b, c \in A$ (associativity)
(2) $a+b = b+a$ (commutativity)
(3) There exists an element—denoted by 0—called the neutral element and having the property $a+0 = a$ for all $a \in A$ (existence of a neutral element)
(4) For every $a \in A$ there exists an element $b \in A$ satisfying $a+b = 0$ (existence of an inverse element)

It is not difficult to show that for an Abelian group $(A, +)$ there exists a unique(!) neutral element—usually denoted by 0—and that every element a has a unique inverse which is typically denoted by $-a$, so that $a + (-a) = 0$.

Superposition produces either 0 or 1 and we light up a lamp on the 2nd floor provided the superposed signal is 1 and similarly for the other floors. There is one obvious question: What should be the image data on the third floor? There are no cables coming in and therefore the lights never turn on.

Now we represent the complete image data in the bold color "orange" among all lighting scenarios.

$$\begin{array}{lcccc} \text{floor 3} & \color{orange}{(0,0)} & (0,1) & (1,0) & (1,1) \\ \text{floor 2} & \color{orange}{(0,0)} & \color{orange}{(0,1)} & (1,0) & (1,1) \\ \text{floor 1} & \color{orange}{(0,0)} & (0,1) & (1,0) & \color{orange}{(1,1)} \\ \text{floor 0} & \color{orange}{(0,0)} & (0,1) & \color{orange}{(1,0)} & (1,1) \end{array} \tag{20}$$

Next we consider the stealthy lighting scenarios and already note that the lack of cables leaving the ground floor means that the whole floor 0 is stealthy; we color the stealthy data "blue" and underline it.

$$\begin{array}{lcccc} \text{floor 3} & \color{blue}{\underline{(0,0)}} & \color{blue}{\underline{(0,1)}} & (1,0) & (1,1) \\ \text{floor 2} & \color{blue}{\underline{(0,0)}} & \color{blue}{\underline{(0,1)}} & (1,0) & (1,1) \\ \text{floor 1} & \color{blue}{\underline{(0,0)}} & (0,1) & (1,0) & \color{blue}{\underline{(1,1)}} \\ \text{floor 0} & \color{blue}{\underline{(0,0)}} & \color{blue}{\underline{(0,1)}} & \color{blue}{\underline{(1,0)}} & \color{blue}{\underline{(1,1)}} \end{array} \tag{21}$$

We note that if data in (20) is orange, then the corresponding data in (21) is blue, but not the other way around.

Data on a specified floor can be added by adding up the coordinates, f.e.

$$(1,0)+(1,1) = (0,1) \quad \text{or} \quad (1,1)+(1,1) = (0,0).$$

We note that on every floor, if we add orange data or if we add blue data, the resulting data is always orange or blue, respectively—this is by design, but perhaps not immediately obvious.

Next we illustrate the procedure of killing the contaminated data on every level. We need to explain the $\ominus$-procedure. For this we view the orange data as "noise" making it difficult to distinguish between two blue pairs on the same floor. To give an example, we pick the blue element $(0,1)$ on floor 0 and add to it all the different orange elements from floor 0 to obtain the set

$$\{\color{blue}{\underline{(0,1)}} + \color{orange}{(0,0)}, \color{blue}{\underline{(0,1)}} + \color{orange}{(1,0)}\} = \color{red}{\underline{\{(0,1),(1,1)\}}}$$

On the right hand side of the above equality, we obtain a set belonging to what we previously called propdata_0.

We can do this for every floor and every blue pair: Namely build a set consisting of all pairs obtained from the blue pair by adding all orange pairs from the same floor to it. Here are the results for the different floors, where we drop the coloring

$$\begin{array}{llll} & \text{propdata on floor 3} & \{(0,0)\} & \{(0,1)\} \\ (22) & \text{propdata on floor 2} & \{(0,0),(0,1)\} & \\ & \text{propdata on floor 1} & \{(0,0),(1,1)\} & \\ & \text{propdata on floor 0} & \{(0,0),(1,0)\} & \{(0,1),(1,1)\} \end{array}$$

There is an interesting feature which has to be pointed out. The collection of elements on each floor allow an addition. For example for the ground floor we define the addition of two sets by building a new set by adding any element from the first to any from the second, for example

$$\begin{aligned} (23) \qquad & \{(0,0),(1,0)\} + \{(0,1),(1,1)\} \\ & = \{(0,0)+(0,1),(0,0)+(1,1),(1,0)+(0,1),(1,0)+(1,1)\} \\ & = \{(0,1),(1,1),(1,1),(0,1)\} \\ & = \{(0,1),(1,1)\}. \end{aligned}$$

or

$$\begin{aligned} \{(0,1),(1,1)\} + \{(0,1),(1,1)\} &= \{(0,0),(1,0),(1,0),(0,0)\} \\ &= \{(0,0),(1,0)\} \end{aligned}$$

We note that whenever we add two pieces of data on a floor the resulting set is already on the list. In other words we have what mathematicians call a group structure, see below. What we just demonstrated works for every floor. We not only have constructed "propdata," but this data has also the structure of an Abelian group.

The above discussion give two examples. For the first example consider A given as the set

$$A = \{(0,0),(0,1),(1,0),(1,1)\}$$

with the addition defined coordinate-wise, and where the coordinate-wise addition is the one in (19). Here the neutral element is $(0,0)$ and every element is inverse to itself, i.e. $-(a,b) = (a,b)$ for $a, b \in \{0,1\}$ since for every $a \in \{0,1\}$ we have that $a + a = 0$.

The second example we have seen is A with the two elements $\{(0,0),(1,0)\}$ and $\{(0,1),(1,1)\}$ and the previously defined addition, see (23). In this case the neutral element is $\{(0,0),(1,0)\}$. The reader will immediately see that this group is just $\mathbb{Z}_2$ in disguise, where instead of 0 we write $\{(0,0),(1,0)\}$ and instead of 1 we write $\{(0,1),(1,1)\}$. The renaming

$$0 \to \{(0,0),(1,0)\} \;\;\&\;\; 1 \to \{(0,1),(1,1)\}$$

defines what is called a group isomorphism: We identify two different sets and this identification is compatible with the additions on the two different sets.

In summary: Given any Morse function $f : \Lambda \to \mathbb{R}$ on a closed smooth space we obtain a directed decorated graph and applying the above magic we obtain a sequence of Abelian groups

$$H_0,\ H_1, \ldots, H_{\dim(\Lambda)}.$$

If we take another Morse function on Λ we obtain again a sequence of groups and—as it turns out, requiring some serious work—they will be naturally isomorphic to the previous ones, i.e. for all practical purposes they are the same. The collection of these groups is called the homology of Λ.

We have a look at another example, namely the previously studied Morse function on the torus, see Figure 8. Note that we have four critical points—one of Morse index 2, two of Morse index 1 and one of Morse index 0. Most important is the observation that there are always two rigid flow lines going from one critical point to another. This immediately implies—an easy exercise—that the images are trivial and that the homology in this case must be given by

$$H_0 = \mathbb{Z}_2, \; H_1 = \mathbb{Z}_2 \oplus \mathbb{Z}_2, \; H_2 = \mathbb{Z}_2.$$

We refer the reader to the appendix on page 141 for a more rigorous treatment of the previous constructions.

CHAPTER 6

Smuggling Ideas

6.1. Arnold's Force

By the 1970s, Vladimir Arnold had made contributions to the canon, [**Ar1**, **Ar2**, **Ar3**], that became the driving force in symplectic geometry. Arnold himself lived as large as his celebrated conjectures. He was known for his superhuman endurance, his penchant for risky adventures (leading to a special rapport with a mountain lion). He was also a dedicated popularizer of mathematics, [**T-Ar**, **M-Ar**].

FIGURE 1. Vladimir L. Arnold.

At the beginning of the 1980s, after having read Arnold's famous textbook, "Mathematical Methods of Classical Mechanics", [**Ar1**], and some of Weinstein's seminal work, [**We1**, **We2**, **We3**, **We4**, **We5**, **We6**]—and before the Conley-Zehnder breakthrough [**CZ**]—Hofer defined symplectic geometry as follows (only for close friends): "the field where one can reformulate any question as a seemingly unrelated new question which still cannot be answered." What has changed since then is that, due to the work starting with Rabinowitz, Conley and Zehnder, Ekeland, Gromov, McDuff, Eliashberg, Floer, and others, many of these questions can be answered. And the fact that "one can reformulate any question as a seemingly unrelated new question" is now an asset.[81] What happened was that the technical machinery came much later than the language—to an extent rarely seen in other mathematical areas. Arnold and Weinstein developed a lot of the symplectic language used today.

We recommend to the reader Weinstein's wonderful short paper [**We5**] providing a historical perspective on symplectic geometry, published in 1981, two years before the Conley-Zehnder theorem. Weinstein writes: "$\cdots$. I think it is not unfair to say that symplectic geometry is of interest today, not so much as a theory in itself, but rather because of a series of remarkable 'transforms' which connect it with various areas of analysis."† The footnote associated to † refers to George Birkhoff's professed fear: a "disturbing secret fear that geometry may ultimately turn out to be no more than the glittering intuitional trappings of analysis" [**Bi**].[82]

Weinstein notes that this may prove "especially appropriate when applied to symplectic geometry."

Meanwhile, symplectic geometry was evolving into a highly developed field with an enormous body of theory. And indeed, today it connects with ever more areas in mathematics and physics.[83] The most powerful tools are associated with the analysis of moduli spaces of pseudoholomorphic curves.

In the fall of 1982, Conley spent his sabbatical abroad in Bochum. For several weeks he and Zehnder visited Moser, who was by then director of the Institute for Mathematical Research (FIM) at the ETH in Zürich. In this setting, various plot lines converge—two mathematical worlds, that of critical point theory and symplectic geometry began a potent interaction.

The direct catalyst was John Mather, a professor at Princeton who was also visiting at the institute. Mather had arrived in Zürich via Paris where he'd just been visiting with his friend Michel Herman. From Herman—via Gromov, via Katok—Mather brought word of an elemental result by Eliashberg, then in Soviet Russia.

Eliashberg began his career in 1972 as a professor at Syktyvkar University in a north-eastern region of the Soviet Union. From the late 1970s, he spent his refusenik years—having been denied permission to emigrate to the United States—as persona non grata in Leningrad. Largely cut off from the mathematical community, he worked as a software engineer. But he'd long been working on the Arnold conjecture, and by 1978 Eliashberg felt confident that he'd obtained a proof in the surface case. He knew that his friend Gromov, then of the State University of New York at Stony Brook, would be interested. Of course, he couldn't send the manuscript by international mail because it would be confiscated by the KGB. Eliashberg needed to convey his goods illicitly. Katok was a willing courier.

6.2. The Eliashberg$\xrightarrow{\text{Katok}}$Gromov Smuggling Operation

Katok was born in Washington, D.C., since his parents were members of the Russian delegation to the Lend-Lease program, part of the American effort to defeat the Axis powers by supplying food, oil and wartime materiel to Great Britain, the Soviet Union, and other Allied nations (see [**Ha**]). After the war, the Katoks returned to Moscow, where "Tolya," as his friends called him, became an influential mathematician in the field of dynamical systems. Famously, Katok had thought about so many things that it was always a good idea to seek his advice if one ran into a seemingly new phenomenon—there was a good probability that "Tolya has been there." He might not have invented a theory, but he'd already cleared the landscape, produced examples or new constructions, and overall a refined understanding.

For Eliashberg, Katok's emigration back to the United States in 1978 presented an opportunity: Katok could deliver Eliashberg's mathematical cargo. The timing, however, was slightly off. Eliashberg wasn't quite finished with his paper. Nonetheless, given the circumstances, he gave Katok the draft. He discovered, about two months later, that in addition to being incomplete, this draft also contained a mistake. Fortunately, Katok was still in Moscow and Eliashberg sent him a new draft by mail. Katok confirmed receipt, and soon thereafter delivered the goods to

FIGURE 2. Yasha Eliashberg $\xrightarrow{\text{AnatoleKatok}}$ Misha Gromov.

Gromov. Gromov then shared the manuscript with the Orsay topology group,[84] who undertook a close study. Alas, there was a mix-up regarding which draft was which—due to an error by either Katok or Gromov—and the group studied the incomplete draft.

The Orsay topology group had quite a number of illustrious members, among them: Michèle Audin, Francis Bonahon, Jean Cerf, Alain Chenciner, Adrien Douady, Albert Fathi, François Laudenbach, Alexis Marin, Valentin Poénaru, as well as Michel Herman.[85] Nevertheless, while reading the manuscript they got stuck—not surprisingly, given the incompleteness and the mistake—and abandoned the project.

Eliashberg, meanwhile, had submitted the paper to Functional Analysis and Its Application—a Russian journal run by none other than Arnold, and Israel Gelfand. One day, Eliashberg called his friend Askold Khovanski, a student of Arnold's. "He told me that he was refereeing my paper and had a few questions," Eliashberg recalled. "We met and talked for a few hours. I answered all his questions to his satisfaction." Khovanski wrote a positive report, and in passing mentioned to Arnold that he'd discussed the paper with the author. "Arnold was furious," said Eliashberg. "He told me that I destroyed his referee"—since the referee's goal, apparently, was to check not only whether the result is correct, but also whether the proof as conveyed was comprehensible. Arnold decreed that the refereeing process had to be restarted.

In 1979, Eliashberg applied for an exit visa from Russia and was refused. He received a call from one of the journal's editors, informing him that while there was no mathematical reason to reject the paper, Gelfand requested that it be withdrawn—"because publishing a paper of such an ideologically bad person like me could be harmful for the journal," Eliashberg recalled. He refused to withdraw the paper, but eventually sent it to VINITI, the All-Russian Institute of Scientific and Technical Information of the Russian Academy of Sciences, where it was refereed in the "Referativny Zhurnal" (Review Journal).

6.3. Mather's Travels

In the academic year 1982-1983, Mather was a visitor at the Institut des Hautes Études Scientifiques, in Bures-sur-Yvette, France. From there, he made the short trip to Paris for a visit with Herman.[86] Herman liked his high-end scotch, and while thinking smoked Gauloise cigarettes. Perhaps it was in such a heady context that he told Mather about the failed attempt at reading Eliashberg's manuscript. It is certainly the case that during the next stop on Mather's travels, with Conley and Zehnder in Zurich, he was interrogated about what had transpired with Herman. As Zehnder remembered [**Ze0**]:

"Michel Herman was the sharpest mind in classical dynamical systems I have ever met, so far. Of course, we inquired immediately what Michel was working on. John Mather told us that Michel together with his group of brilliant students tried very hard, and in the end not successfully, to decipher an unreadable and un-understandable manuscript of an unknown Russian mathematician named Eliashberg, about the Arnold conjecture for surfaces, in particular the 2-torus. So, to be polite, Charley asked him what this conjecture was."

FIGURE 3. John Mather and Michel Herman.

Mather explained the conjecture. Conley and Zehnder looked at each other and declared that they could provide a proof for the torus in all dimensions $2n$ in a few days. When they delivered, Mather said: "You are very fast in proving old conjectures. How about proving another old conjecture, the Riemann Hypothesis?"

"Again, we asked what that conjecture was," said Zehnder. "But then, obviously, did not touch it."

6.4. The Conley-Zehnder Theorem

Here is the gist of the Conley-Zehnder contribution: Starting with the standard symplectic vector space $(\mathbb{R}^{2n}, \sum_{i=1}^{n} dq_i \wedge dp_i)$, we call two points $x, y \in \mathbb{R}^{2n}$ equivalent provided their difference $y - x$ belongs to the integer lattice $\mathbb{Z}^{2n} \subset \mathbb{R}^{2n}$. The set of all equivalence classes $[x]$ can be identified with the $2n$-dimensional torus

$$T^{2n} = S^1 \times \ldots \times S^1,$$

and moreover it inherits the structure of a symplectic space.

Given a time-dependent Hamiltonian H, which we may assume to be 1-periodic in time (due to the local structure of the symplectic group discussed on page 139),

$$H : S^1 \times T^{2n} \to \mathbb{R}$$

we obtain the time-dependent Hamiltonian system

$$\dot{z} = X_{H_t}(z) \tag{24}$$

whose time-1-map is called a Hamiltonian diffeomorphism.

The famous Conley-Zehnder result, achieved in 1983, is given by the following theorem [**CZ**]. In the case of the torus, the estimates coming from the Arnold conjecture are optimal—there are examples attaining the predicted minimal number of periodic orbits.[87]

THEOREM 6.4.1 (Conley-Zehnder). *For the standard torus (T^{2n}, ω) a Hamiltonian diffeomorphism Φ has at least 2^{2n} fixed points if they are non-degenerate and in the general case at least $2n + 1$ fixed points.*[88]

As Zehnder recalled, it earned a great compliment from Gromov: "This result is so beautiful, I would not mind stealing it!"[89] The Conley-Zehnder breakthrough was quickly followed by several results exploring the ideas further [**Ch2**, **Ch1**, **CZ1**, **Fo**, **FW**, **H1**, **LS**, **F5**, **Si**]. Noteworthy is [**Ch1**] which, analyzing [**Ch2**], introduces the appropriate version of the "theory of generating functions" which later leads to further advances and ultimately to the use of sheaf-theoretic methods (in certain situations), a topic of current interest.

FIGURE 4. Nancy Hingston.

The Conley-Zehnder paper [**CZ1**] is the starting point for the study of periodic points of iterated symplectic maps leading to an influential conjecture by Conley. In 2009 Nancy Hingston, [**Hi**], solved the conjecture for tori with a very sophisticated argument. This started a whole development concerned with questions about the number of periodic solutions of Hamiltonian systems or the number of periodic points of symplectic maps.

Part 3

Breakthrough

CHAPTER 7

Floer Between Berkeley, Bochum, and Beyond

During his first year at Berkeley, Andreas lived in Cunningham Hall, a typical university dorm and not quite to his taste. In the second year, he moved into Barrington Hall, a student housing cooperative. Murals and graffiti captured the ethos: "Please keep arms and hands inside the ride at all times." And, "Better living through chemistry." The drug scene was no secret, and Floer partook. He reported to friends that so-called "chemists" designed just the right drugs for upcoming parties and concerts, and that the local group Grateful Dead were especially effective in stimulating chemical creativity. Police were known to raid the hall, and on at least one occasion they physically carried students out—but then released them, whereupon the students re-entered and were carried out again. Eventually, the police made arrests, including Andreas.

One of his closest friends at Barrington Hall was the American Rebecca Steinberg, now an expat in Berlin. Steinberg was a house manager who studied politics and music. Together she and Andreas listened to a lot of Bach. "Andreas blossomed when he heard music that he loved, he got this look of serenity," Steinberg said. Even in his steady state, Andreas at times seemed ethereal. "He would float with the wind," she recalled. He was "wispy," a lean $6'2''$ with very long brown hair (later kept closely shorn). "He didn't talk a whole lot, he was very soft-spoken. But he would smile." She considered him a lonely soul, emotional, sensitive to little things, loyal like a puppy. "If you were his friend, you were really his friend," she said. "He was honest and forthright and trusting. He was a kind and loving person. He really lived on a different plane."

They met through Leo Sullivan, a talented pianist and Steinberg's flatmate. Andreas was drawn in by the sense of family that Leo built among a group of friends, with music being the common factor. Andreas played the viola da gamba. But mostly he loved to listen, astutely. He relished Mozart, Beethoven, Wagner and Verdi; piano, chamber, symphonies, concertos, sacred, choral, and twentieth-century.[90] Sullivan possessed a vast repertoire of musical knowledge. He DJed listening parties, spinning records of favorite pieces and composers—as Steinberg put it, Leo dealt music before he dealt drugs.

Leo cultivated himself as a shaman-like character who talked about drugs as magical and mystical, an entrance to an otherworldly realm. Steinberg, who shared in the drugs, if infrequently, remembered once walking through the flat and finding Andreas and Leo, and Leo's girlfriend Hélène McGiffin (who then went by "Phread") in the midst of a trip: They were talking about how they could see the structure of the building lit up as neon rods, colors emanating from the walls. "It was as though they were discovering a new world," she said.

At the time, there was a palpable sense of nostalgia: For Berkeley of the 1960s—there were still resounding echoes of the free speech movement, Mario Savio's sit-in address on the steps of Sproul Hall. For Buffalo Springfield's "There's something happening here/What it is ain't exactly clear." For the Beat poets, William Burroughs and Allen Ginsberg[91] and Jack Kerouac. "There was this feeling of being on the cutting edge, doing something other people hadn't done and experiencing something new," Steinberg said. It started with hallucinogens, and seemed innocent and fun. Then it evolved to harder drugs, coke and heroin, and got ugly and scary.

According to available documentation, Andreas lived at Barrington Hall for three terms: fall 1983, spring 1984, and fall 1985. The extent of his drug use is difficult to definitively determine. Kaiser pieced together arguably the most vivid picture of Andreas' life from his letters and visits back to Bochum. By all indications, Andreas took something of an experimental approach not only to drugs, but also sexuality. Although regarding the latter, he seemed indifferent; he could have done without any appreciable sex life. Nonetheless, Andreas suffered unrequited love. Kaiser heard a lot about Leo: "Every third word was 'Leo'," he recalled. "It almost annoyed me." Andreas was smitten with Leo for a time, but it was likely both undeclared and unnoticed. And this was not Andreas' only "unilateral romance." There had been previous objects of such affection, and others followed, men and women both. A number of years later, Andreas proposed to Hélène after she and Leo had broken up.

On the matter of drugs, however, Andreas was more secretive, and delusional. "He was—everyone agrees—a mastermind and so he mastered everything with his mind," Kaiser said. "And he felt positively immune to drugs. He thought he could ban addiction with the wonderful power of his mind. That was his hubris"—the fool does not know his own limits.

Some people in his mathematical circle noticed the effects. Rob Kusner, then at Berkeley doing his Ph.D. in mathematics, now at University of Massachusetts at Amherst, recalled the time when Andreas accepted an invitation to join a group carpooling for a daytrip to Stinson Beach. Early one September morning, the station wagon, all but full, arrived to pick up Floer. There was no sign of him. Kusner didn't know him well, and wasn't sure what to do. After waiting a while, he knocked softly on the door; no response. He knocked a little harder; still no response. Then he banged on the door, in the style of the "blue meanies"—Floer emerged, looking a bit wrecked, essentially asleep. He climbed aboard, but declined to chip in for gas money. "He refused, really, on principle," Kusner said. "There was no marginal cost for taking him."

Over the years, nobody saw much of Floer. He was like a ghost; he floated through the scene every now and then. He wasn't taking any classes. He was clearly working hard on something. Kusner was impressed, and intimidated, especially by Floer's jam-packed standing-room-only presentation at the Friday Geometry Seminar in the basement of Evans Hall (Kusner was scheduled to speak a few weeks hence). Floer was clearly on an impressive trajectory, said Kusner, might be even destined for greatness. But Kusner also worried that Floer might be doing himself in. Floer would hardly be alone in such an endeavor at Berkeley. "A lot of people came here with great ambition, and they burned brightly," said Kusner. "And then a lot of them just burned out."

He added: "Andreas was an extraordinary person—exceptional. That's why you're writing a book about him."[92]

* * *

Mathematically, Floer's fearlessness, his anti-authoritarian streak, served him well. Both are important when pushing against the current of conventional wisdom. In science, if one navigates near the boundary of knowledge, things become murky. Opportunity at the boundary is abundant, yet to reap the rewards one must be comfortable with discomfort, frustration and uncertainty. Floer had this mindset.

As a grad student participating in the Donaldson seminar, Floer had learned about gauge theory and bubbling-off analysis. Soon thereafter, he began collaborating with Weinstein on nonlinear Schrödinger equations.[93] Given this collaboration, it seemed obvious to some that Weinstein was Floer's advisor. But Floer had also started working with Taubes, who gave him a problem about monopoles on asymptotically flat manifolds, which Taubes had tried unsuccessfully. Given this work, others assumed that Floer's advisor was Taubes. Taubes, when asked if he had been Floer's advisor, immediately answered that it was Weinstein. Asking Weinstein resulted in a "limit cycle" since he immediately answered that it was Taubes. Berkeley mentoring at its finest—Floer was the Ph.D. student you never knew you had.

7.1. With Loving Care

While abroad, Floer kept in touch with friends in the Ruhrgebiet and mathematicians at the Ruhr-Universität Bochum. Ultimately it was from Bochum that he received his Ph.D.—a curious turn of events that has generated speculation. He had avoided the German military service by doing a so-called Ersatzdienst—a sort of community service for conscientious objectors—at the Technisches Hilfswerk (the federal civil protection organization). This allowed him to spread the service over a number of years, spending summers in Germany. Having started in Berkeley in the fall of 1982, he likely would have been back in Germany for brief periods during the spring and summer of 1983. When he visited the university, he was given an office, and he began an ongoing conversation with Zehnder.

"One day a student knocked on my office door on the 7th floor," Zehnder recalled. "I had never seen him before in my analysis lectures. He did not introduce himself but just asked, 'Do you have an interesting topic for my thesis?' " Zehnder explained the Arnold conjecture and described his proof with Conley. For Floer's thesis, he proposed: Analyze the structure of bounded solutions of the artificial gradient flow, using Conley's index theory. "He was immediately attracted to the challenging problem," Zehnder said. "I suggested that he should start with the Arnold conjectures for surfaces of higher genus in order to verify the claims of Eliashberg."[94]

When back in Berkeley, Floer often called Zehnder to consult—always at midnight in Bochum. Zehnder observed that while Floer's ideas were beautiful, his notes were a mess. "So I told him to write his ideas down 'with loving care, for the benefit of the reader.' At first, he was furious, but later on he always made fun of me with 'my loving care writing style.' Whenever he showed up with new work, Andreas announced: 'Herr Zehnder, Sie können mich jetzt wieder in der Luft zerreissen'—now you can tear me apart again."[95]

Floer's progress on the Arnold conjecture for surfaces required that he exhibit some refinement of the Conley index; namely, prove a topological continuation theorem for normally hyperbolic invariant sets, [**F6**]. This result then provided a key input to the proof of the Arnold conjecture for surfaces and certain Kähler manifolds, [**F5**]. These two papers, [**F5**, **F6**], were submitted to journals in early 1985. Floer had submitted his Ph.D. thesis (roughly the same material) the previous December and a preprint circulated as well.

His friend Kaiser recalled that Floer had spent the fall of 1984 in Bochum. Floer told Kaiser that his Ph.D. was written quickly based on shelved material (likely from summer 1983). Kaiser speculated that Floer needed a certain amount of hours to finish his Ersatzdienst service, and during the same period wrote up his thesis.

A more intriguing explanation for Floer's Bochum Ph.D. is due to Bahri. Bahri speculated that the speedy Bochum Ph.D. might have been motivated as follows: There were several Arnold conjectures, and one of them was about exact Lagrangian embeddings into cotangent bundles. After the Conley-Zehnder result, Marc Chaperon proved the latter in the case of the torus,[**Ch2**]; and Hofer proved the general case in June of 1984, publishing a year later in [**H1**]. At that time Hofer had some joint projects with Ekeland, and he told Ekeland about his general result. Bahri had also indirectly learned of it (possibly through Ekeland), and Bahri in turn told Floer. This news might have panicked Floer—he might have assumed it was the same Arnold conjecture that he was working on. Perhaps he then quickly wrote up his ideas, turning the shelved material into a Ph.D. thesis.

Floer, however, had never mentioned anything along these lines. Thus Bahri's speculation, while tantalizing, was likely wrong. Kaiser's more pragmatic account was likely correct.[96]

As Floer completed this work, [**F5**], François Laudenbach's graduate student, Jean-Claude Sikorav, was also finishing a paper on the Arnold conjecture inspired by the Conley-Zehnder theorem. Sikorav's paper covered certain closed symplectic manifolds, admitting metrics of non-positive sectional curvature and some additional technical condition which are always satisfied by closed Riemann surfaces [**Si**]. Both Sikorav and Floer acknowledged each other's independent work in their publications.

7.2. The Gromov Paper—A Short History

In January 1985, when Hofer delivered a talk about the Arnold conjecture for exact Lagrangian embeddings in Paris, he learned about Gromov's breakthrough work on pseudoholomorphic curves, which just was circulating as a preprint [**G**]. Sikorav, in the audience, shared a copy with Hofer. On the way home, Hofer, then a lecturer at the University of Bath, skimmed through (tough reading). Having the background of Rabinowitz's breakthrough, the saddle point reduction in [**Am**, **AZ**], the Conley-Zehnder result [**CZ**], and his own work [**H1**, **H2**], it immediately crossed his mind that rather than studying a flow associated with the bad variational problems, perhaps one could study just a partial differential equation. This was similar to what had crossed Zehnder's mind, but augmented by the insight from Gromov that the action functional without a Hamiltonian perturbation can be studied. Nevertheless, it was clearly a very tall order.

Gromov had earned a reputation for being wild—intellectually and otherwise—despite typically wearing his shirt tightly tucked. He had learned about the Arnold conjectures during a visit by Dimitry Fuchs in Moscow. His initial reaction was that the conjectures were all wrong. He set out to disprove them, and failed. But in discussions with Eliashberg, they figured out that there was a "conjectural borderline, either this or that"—the soft versus hard alternative in symplectic geometry.[97]

Gromov always thought that there should not be a theory on the hard side (everything is governed by an h-principle, see [**G1**].). Whereas Eliashberg constantly changed his mind.[98]

FIGURE 1. Floer circa 1986.

We know today that the symplectic landscape includes both: the soft problems governed by h-principles, and hard problems with obstructions that arise from, for example, pseudoholomorphic curve theory (including Floer theory, of course).[99]

As an aside, much of symplectic geometry is about mapping this landscape, and it seems that there is not much room between the hard and soft regions. It is therefore an interesting question to understand where the soft-hard transition occurs: Is it abrupt or gradual? Surely something can be learned there! Indeed, Emmy Murphy's surprising 2012 Ph.D. results showed that in higher dimensions there is very little room between the soft and the hard—Murphy's work strongly indicates the abrupt option. For this and other work she was awarded a 2020 New Horizons Prize.

FIGURE 2. Emmy Murphy.

The impact of Gromov's paper was tremendous—it was mentioned in the citation for his 2009 Abel Prize. It contained an enormous amount of material in a terse style. For Floer, Gromov's pseudoholomorphic curve theory provided a technological advance, a tool, that bid "good-bye" to flows—replacing them by elliptic partial differential equations with asymptotic boundary conditions. These elliptic differential equations are symmetry breaking perturbations of the pseudoholomorphic curve equations. Gromov, hearing about Floer's work, initially dismissed it.

However, Gromov held in high regard Dusa McDuff's contribution, which in a sense started the systematic hunt for pseudoholomorphic curves. McDuff, having worked on operator algebras and diffeomorphism groups, became interested in the "soft side" of symplectic geometry a few years before Gromov's paper [**G**] appeared—the big message of that paper was the existence of the "hard side." She immediately started working on the hard side, and clarified Gromov's ideas in a series of papers. She gave the first example of symplectic forms on a compact manifold which are cohomologous but not diffeomorphic [**McD1**] using Gromov's pseudoholomorphic curve theory.

FIGURE 3. Dusa McDuff.

This work was followed by a deep classification result for rational and ruled symplectic 4-manifolds [**McD2**], showing that symplectic methods have something nontrivial to say in topology. McDuff also did important work with François Lalonde, [**LMc2**, **LMc3**] on the classification of symplectic 4-manifolds; and on "Symplectic Energy" and "L^∞-Geometry", [**LMc1**,**LMc4**]. These works were motived by Hofer's paper [**H8**] which starts what is now called "Hofer's Geometry." Leonid Polterovich developed these ideas further resulting later in the important book [**Pol**].

McDuff and Dietmar Salamon, to the benefit of the emerging field, wrote a number of books on symplectic topology as well as pseudoholomorphic curves. For their influential book on "J-holomorphic Curves and Symplectic Topology" they received the 2017 Steele Prize for Exposition from the American Mathematical Society.

McDuff, at an early stage of the symplectic venture, had shown her sustained ability to produce amazing examples, which have guided the development of the

FIGURE 4. Dietmar Salamon, François Lalonde and Leonid Polterovich.

field. Later, after Floer had developed his fundamental ideas, he told Hofer that he believed the Liusternik-Shnirelman version of the Arnold conjectures to be wrong in full generality, and that counterexamples should exist. But he thought it was above his pay grade to find such examples. He suggested the only possible way to find them: "We just have to get Dusa excited about this problem." The general Liusternik-Shnirelman case—that is, the fixed points of the Hamiltonian map are allowed to be degenerate—is still not solved in full generality. McDuff was too busy (dis-)proving other things.

7.3. Zeroing in on Floer Homology

In spring 1985, when a preprint of Gromov's work arrived in Berkeley, Weinstein organized a seminar. Floer was then back at Berkeley, where he continued life as a grad student even after having received his Bochum Ph.D. He briefly occupied his own barely furnished apartment. In this setting, he had his big breakthrough: Gromov's input melded with ideas already percolating about Conley's index theory, the Conley-Zehnder theorem, and the Witten-type approach to Morse theory.[100] Floer had a vision, modulo the technical details. He had stumbled upon the hidden door in the jungle and swung it open.

Naturally, with this vista on the horizon, Floer was moved to share his excitement. But what in a Hollywood movie would have been a moment of triumph, with a bursting symphony soundtrack, became in reality a rather unceremonious affair. Floer knocked at Taubes' door and declared: "Now I can do infinite-dimensional Morse theory!" Taubes was busy packing up his office for a move to Harvard, and shooed him away. "I kicked him out of the office," said Taubes. Taubes added that had he a dime for every time he should've been impressed but wasn't, "I would be a rich man now." Floer didn't press the matter; he obligingly left.

From 1985 to 1988, Floer faced the arduous task of getting the technical details down on paper. While he joked about all the preprint requests he was receiving—starting in the summer of 1986—for the papers he'd not yet written, it was apparent that the situation had him on edge.

He had applied for postdoc positions and accepted an instructorship at the State University of New York at Stony Brook (SUNY), where he began in the fall of 1985. Before leaving Berkeley, he finished the write-up of his work about the Arnold conjecture on surfaces—it added some new ideas to the proof of the Conley-Zehnder theorem, [**F5**, **F6**]. One of the new results was a global topological continuation theorem for normally hyperbolic invariant sets. He also wrote a survey article with Zehnder describing the status of the work on the Arnold conjectures, [**FZ**], and completed the paper with Weinstein, [**FW**]. And at about this time he started to understand the bigger picture about the Arnold conjecture and its main ingredient—what became Floer homology, a.k.a. Floer theory—i.e., that which he had tried to convey to Taubes. While at Stony Brook, Floer solved Taubes' hard problem, eventually resulting in two papers [**F8**, **F9**] concerned with monopoles on asymptotically Euclidean three-dimensional manifolds. In July 1986, he submitted [**F7**] a research announcement to the Bulletin of the American Mathematical Society summarizing the content of these papers, and indicating they would appear in the Communications of Mathematical Physics. They never did.

For one paper, "Monopoles on asymptotically flat manifolds," Floer had received a first referee report saying something to the effect of: 'Great result, atrocious writing.' Floer provided a revision, which elicited a similar sentiment in a six-page report, which began:

"This paper is an improvement over the previous version; however, there are still several problems which should be resolved before the paper is published. Careless writing, especially in the introduction, leaves confusion in the reader's mind as to just what will be proved, and how. There are numerous inconsistent statements. Notation is used before it's defined and is inconsistent from line to line in a few places. Many of the unclear points are minor, individually. However, they are so numerous that many theorems (statements as well as proofs) become almost unintelligible. For each unclear point the reader may be 95% clear what is meant; after twenty of these his compounded uncertainty is rather large."

The paper never underwent a second revision. Both manuscripts remained unfinished upon Floer's death. Yet they contained sufficient detail to convince the reader that he had solved Taubes' problem. Perhaps Floer had postponed work on those papers to focus on developing Floer theory.

As Floer worked out details of the Arnold conjecture for more general cases, he began to feel at ease with giving talks about the subject. But he wasn't happy in the isolated Stony Brook setting. His closest friends—mostly non-mathematicians—were in California, France, and Germany. He applied again for postdoctoral positions. Now, with no shortage of good results on his curriculum vitae, he was more competitive. Zehnder, writing a letter of recommendation, mentioned that Floer had a breakthrough on the Arnold conjecture in the pipeline.[101] Floer received an offer from NYU's illustrious Courant Institute, which he was pleased to accept. He arrived in NYC for the fall of 1986.[102]

CHAPTER 8

World Premiere

8.1. The Arbeitstagung

That summer, Floer had visited Bochum and made a short trip south to Bonn for the Arbeitstagung, an annual international "working conference" for the dissemination of new ideas, founded by Friedrich Hirzebruch. A number of the speakers were chosen in advance, while others were determined by something of an ad hoc democratic process: The audience shouted names of desired speakers, and Hirzebruch, together with Sir Michael Atiyah, a de facto co-organizer,[103] composed the balance of the program according to their selective hearing.

FIGURE 1. Sir Michael Atiyah and Friedrich Hirzebruch.

Understanding this modus operandi, Floer talked with Atiyah beforehand and pitched his recent work. It is not known whether Atiyah or Hirzebruch had already heard of his achievements, but Floer made it onto the program, see Appendix B. The Arbeitstagung marked his first public lecture about Floer homology.[104] The meeting also occasioned some choice advice from Atiyah. He suggested that Floer study the Chern-Simons functional[105]—one of those functionals that one should not have been able to study. The big picture was (maybe not yet in this precise form) that the upcoming Donaldson invariants, which were invariants for smooth structures on closed four-manifolds, should have a relative version for compact smooth four-manifolds with boundaries. In this case, the invariant should take its values in a Floer-type construction associated with a three-manifold, i.e. its boundary.[106]

8.2. Collaborative Jackpots

Fortified with that gem from Atiyah, Floer arrived at the Courant Institute in the fall of 1986. The French mathematician Claude Viterbo, who'd just received his Ph.D. with Ivar Ekeland, started there at the same time. In France, it was possible to fulfill one's military service obligations by assisting (in the field of education and research) a developing country—the French military exhibited an unexpected amount of humor in allowing Viterbo to spend his military service in New York City.

FIGURE 2. Claude Viterbo

At Courant, Viterbo made the first breakthrough on the Weinstein conjecture. He showed that a regular compact hypersurface of contact-type in $\mathbb{R}^{2n}$ satisfies the Weinstein conjecture, [**Vi1**]. Viterbo used more classical techniques from his Ph.D. work and a trick that allowed him to use methods known from convex analysis in a non-convex setting. This ground-breaking paper led Hofer and Zehnder to an even more general result, introducing the "almost existence phenomenon," which trivially implied the Weinstein conjecture for all compact regular hypersurfaces of contact-type in $\mathbb{R}^{2n}$.

FIGURE 3. Eduard Zehnder and Helmut Hofer.

This motivated a very short paper (really more of a note) that was essentially obtained overnight, [**HZ1**]. "I remember well that we hit such a jackpot in a single night at a conference in a creative burst, with great enthusiasm and with lots of beer," Zehnder once recounted. "With an all-round attack we had completely killed a problem." This result [**HZ1**] established an idea that is still used when thinking about the Weinstein conjecture in higher dimensions.

The conference in question was that NATO Conference at Il Ciocco, Italy, in October, 1986, organized by Paul Rabinowitz jointly with Antonio Ambrosetti, Ivar Ekeland, and Zehnder, [**RAEZ**].[107]

On this occasion in Il Ciocco Zehnder simultaneously experienced not only a great high (writing a maximal paper in minimal time with a critical amount of beer), but also a memorable low: In transmitting Floer's startling new work, he had elicited Gromov's wrath—his disbelief that anything good could come from Floer's symmetry breaking.

Around the same time, Floer, Hofer, and Viterbo began collaborating. Their joint work, [**FHV**], seemed to be the first paper using Gromov's pseudoholomorphic curve theory in Floer's modified form (breaking the symmetry of the pseudoholomorphic curve equation), to attack the Weinstein conjecture on more general spaces. It implemented the idea of almost existence from [**HZ1**] on more general spaces. This idea was then generalized by Hofer and Viterbo in [**HV**], before Gromov-Witten theory was even formalized, showing that a sufficiently rich structure of moduli spaces of pseudoholomorphic curves implied the Weinstein conjecture. Later, after Gromov-Witten theory was established, Liu and Tian, [**LT**], used the set-up developed in [**FHV**, **HV**] to prove a more definitive result. It showed that a sufficiently structured Gromov-Witten theory for a closed symplectic manifold implies that its contact-type compact regular energy surfaces satisfy the Weinstein conjecture.

8.3. The Importance of an Atomic Clock

While at Courant, Floer was protective of his time. He did make room in his schedule for Hofer and Viterbo, if within certain parameters. Since Hofer was then at Rutgers, he would travel to Courant one week, and then the following week Viterbo and Floer would do the reverse trip. As one of the meetings in New York concluded, Hofer still had a question and he asked Andreas if they could briefly discuss. After inquiring about the hour, Andreas proclaimed: "Courant is paying me only to do mathematics from 9am to 5pm." He kept to a strict closing time.

Rolf Kaiser also witnessed Floer's ambivalence about the notion that professional mathematics demanded the whole person, and to a great extent the whole of their time. "He never wanted to be like that," said Kaiser. He treasured his friends who didn't talk shop. Then again, Andreas was known to get distracted by a Geistesblitz—a "brainstorm"—at inopportune moments. "I'll never forget the view of him sitting, brooding, and obscenely high-concentrated [on math] under dancing disco-lights... When he wanted to solve a problem, he was pondering and pondering, no matter when or where, until it came." After his mother of all Geistesblitzes, Andreas, in his inimitable combination of pride and modesty, "knew immediately what it was that he had found," Kaiser said—and knew what he was worth.

FIGURE 4. Louis Nirenberg.

In 1987, Floer was asked to give a talk in the Courant-Princeton-Yale analysis seminar. It was held at Courant, and since it took some effort to get there from Princeton or New Haven, the meeting was held once a month, featuring two talks, each only 45-minutes long. With the congregants gathered in the lecture theatre at Warren Weaver Hall, Percy Deift introduced Floer and also reminded him that he should finish his talk on time because the second speaker had to catch a plane.

Yet on this particular day, several factors were not in Floer's favor. For one, he mistakenly assumed it was a standard 60-minute lecture. In addition, unbeknownst to Floer, the large clock on the wall of the lecture theater was broken; and he never wore a watch. For a while, he proceeded without trouble, concentrating intently. At some point, however, he began checking the clock on the wall, which informed him that time was passing very slowly. He realized that he would soon run out of material.

Improvising, he inserted meandering philosophical tangents. At the 50-minute mark, the chair interrupted and told Floer that he was already over time and that he should wrap up. Floer asked how long he'd been talking; 50 minutes said the chair. Floer pointed to the clock: "This cannot be, I still have twenty minutes remaining." Deift informed him that this clock was broken. Agitated, Floer addressed his audience: "I can't understand you people. How can you give lectures in lecture theaters where the clocks don't work? We in Germany have atomic clocks in the lecture theaters." While the audience was still trying to figure out if Floer had made a joke, Louis Nirenberg, sitting in the front row, and as quick-witted as ever, countered with a question: "But Andreas, is there not the danger that the German students get radiated?" The audience laughed; Floer was unmoved (more precisely, he was pissed). During a short break, Viterbo, possessing the sustained ability to say the right thing[108] at the right time, induced Floer to laugh at the absurdity.[109] Floer recovered, and the second speaker caught his flight.

CHAPTER 9

Getting Inside Floer's Head

When up-and-coming mathematicians learn about Floer theory, the material is embedded within new viewpoints and future applications, and the narrative looks very natural within that landscape. But this pre-digested view is devoid of historical perspective. It is also striking to note that the cartoon pictures typically used to describe Floer theory are quite easy compared to what's actually needed for rigorous implementation. "It's mind-boggling to see it work," said Akshay Venkatesh, who recently began contemplating shadows of Floer-type structures in number theory. "There is such a difference between the beautiful cartoon picture and the staggering amount of work in analysis that you have to do in order to make that picture take mathematical shape. Floer extracted this very rigid, nice structure out of what seems to be an untamable mess."

In what follows, we attempt to reconstruct the raw ins and outs of Floer's thinking, sketching some of the complexities as they coalesced—and ultimately led him to proclaim: "Now I can do infinite-dimensional Morse theory."

Floer's initial thinking was, of course, motivated by the Conley-Zehnder theorem and the Conley index theory. This work concerned the Arnold conjecture for standard symplectic tori and the Conley-Zehnder framework allowed a simplified approach to the problem that was not possible in more general cases. Conley and Zehnder found an exciting part of the jungle and noticed a small door. Their theorem illuminated this door like light through the jungle canopy—and suggested that something of intrigue lay beyond. They investigated the possibilities, thinking about more general cases. As Zehnder once said: "As you all know, all these pioneering results have been extended later on to more general symplectic manifolds. Of course ... we did try to apply our ideas to general compact symplectic manifolds. We wrote down on $\mathbb{R}^{2n}$ the equation of a bounded orbit of the non-regularized L^2 gradient flow of the action functional and ended up with a non-linear asymptotic boundary value problem for an elliptic system of PDEs, but did not know how to proceed from there," [**Ze0**]

The available mathematical tools seemed useless, and there were no hints about what new tools to invent. Nevertheless some of the methods—that Floer had picked up around 1985 from Conley, Zehnder, and Taubes—turned out to be useful in Floer's joint work with Weinstein on the non-spreading of wave packets for a cubic Schrödinger equation, [**FW**].

Opening the Conley-Zehnder door seemed to require not an old-fashioned key but rather an RSA-caliber code. One must listen to the problem—poke at it and note the response. Does it resonate with what one knows? Or if not, how can one breach the void?

9.1. Starting with the Conley-Zehnder Proof

When Conley and Zehnder learned from Mather about the Arnold conjecture they immediately knew that they could prove this conjecture for tori. The underlying reason was that they had spent considerable time studying a technically more advanced problem.[110] The study of the problems Conley and Zehnder were occupied with had started with the Amann-Zehnder collaboration, [**AZ**, **AZ1**] based on [**Am**], and was concerned with the understanding of Hamiltonian systems on $\mathbb{R}^{2n}$ having different behaviors near a rest point and infinity. The change of this behavior, which can be quantified by what is now called the Conley-Zehnder index, should under suitable conditions guarantee the existence of nontrivial periodic orbits ([**CZ2**, **CZ3**]). Their ideas originated from the attempt to generalize the Poincaré-Birkhoff twist theorem, a well-known result in dimension two, which says that an area-preserving self-map of a closed annulus twisting the two boundaries in different directions needs to have a fixed point—at least two, in fact (see [**Bi2**]). The Poincaré-Birkhoff theorem has been well studied and generalized, particularly by John Franks and Patrice Le Calvez (see, for example, [**Fra1**, **Fra2**, **FrLC**]).

In the case of the Arnold conjecture for the torus, one is given a smooth Hamiltonian

$$H : S^1 \times T^{2n} \to \mathbb{R}$$

and is interested in 1-periodic orbits of the system of ordinary differential equations

$$\dot{q}_k(t) = \frac{\partial H}{\partial p_k}(t, q_1, p_1, \dots, q_n, p_n) \quad \& \quad \dot{p}_k(t) = -\frac{\partial H}{\partial q_k}(t, q_1, p_1, \dots, q_n, p_n) \tag{25}$$

where k varies in $\{1, \dots, n\}$. Before Rabinowitz, it wasn't clear how to even attempt to find periodic orbits for this system.

Of course, Rabinowitz' breakthrough changed things. The work by Amann and Zehnder provided the tools to immediately see that finding 1-periodic orbits for (25) is equivalent to a finite-dimensional critical point problem—even if they didn't consider this equation.

After the work with Zehnder, Amann shifted his interests to reaction-diffusion equations, while Zehnder started a collaboration with Conley with the purpose of generalize the famous Poincaré-Birkhoff theorem to higher dimensions.

The fruitful (random) meeting with Mather made Conley and Zehnder aware of the Arnold conjecture, which in the torus case is the problem of finding 1-periodic solutions of the system (25). However, Conley and Zehnder know that such solutions can be found, at least formally, as critical points of an associated variational principle which is defined on an infinite-dimensional space—the action principle on T^{2n} perturbed by the averaged energy along a (contractible) loop

$$z \to \mathcal{A}_{T^{2n}}(z) - \int_{S^1} H(t, z(t))dt.$$

From their previous work it is clear that one can carry out a global reduction, [**Am**, **AZ**], to an equivalent finite-dimensional critical point problem. The new variational problem is of the form:

$$\Phi_H(\bar{r}, x, y) = -\frac{|x|^2}{2} + \frac{|y|^2}{2} - b(\bar{r}, x, y), \; (\bar{r}, (x, y)) \in T^{2n} \times (\mathbb{R}^N \times \mathbb{R}^N),$$

with a bounded b having also a bounded gradient ∇b. The dimension of the domain of Φ_H depends on the integer $N >> 0$, reflecting the size of H. Also, if N is large enough we can homotope Φ_H by perturbing b away to the functional

$$\Phi_0 : T^{2n} \times \mathbb{R}^N \times \mathbb{R}^N \ni (\bar{r}, x, y) \to -\frac{|x|^2}{2} + \frac{|y|^2}{2}.$$

For the latter, the Conley index theory can be computed and by homotopy invariance it has to coincide with that of Φ_H. Since the Conley-index is essentially determined by T^{2n} it is clear that the Arnold Conjecture has to hold. This was essentially the Conley-Zehnder split-second reasoning.

Conley's theory is generally concerned with the study of flows, particularly their closed invariant subsets and associates an index to an isolated closed invariant subset of a flow. This index is the homotopy type of a certain (cooked-up) space—its Conley index. The power of the Conley index theory is that the index does not change when we deform the flow. Or, more mathematically, this holds when we homotope the flow, as long as we can suitably control the compactness of the isolated closed invariant subsets during such a deformation.

Why is this useful? The answer is simple: studying an unknown flow we might be able to deform it to a simpler one for which we can actually compute the Conley index. This is precisely the case in the situation that must be studied for establishing the Conley-Zehnder theorem. The gradient flow associated to Φ_H is complicated and the gradient flow for Φ_0 (the case when $H = 0$) is rather trivial. Fortunately one can deform Φ_H to Φ_0 in a way that does not change the index. As it turns out, that is enough to conclude from the properties of the Conley index theory the desired result, see [**CZ**]. Not to create the wrong impression: The procedure is nontrivial due to the various technical points which have to be carefully resolved, but from a bird's eye view it follows a very clear path.

Hence, following Mather's missive, Conley and Zehnder realize, essentially in a split second,[111] and with high confidence, a way to deal with the problem. The slightly more painful part was to write it down, [**CZ**].

In his Ph.D. work, Floer generalized the Conley-Zehnder result to surfaces and certain Kähler manifolds, [**F5**]. The ideas he used remain within the Conley-Zehnder framework, but require one to deal with harder technical questions and an extension of some of Conley's ideas, see [**F6**]. At this stage, Floer was still toiling at the threshold of the Conley-Zehnder door.

Ideas leading to solutions of longstanding problems usually originate outside the prescribed field. In 1983, the use of flows to study variational problems was ingrained in Conley and Zehnder's realm. Previously, successful activity had started with the work of Ambrosetti and Rabinowitz,[**AR**]: using variational methods for the study of nonlinear partial differential equations.[112] In all the literature—that is, the "cultural noise"—the vibe was:

FLOWS, FLOWS, FLOWS.

Even if Conley's view was different and novel at the time, it operated under a similar premise:

FLOWS, FLOWS, AND EVEN MORE FLOWS.

There was, however, an important difference. Ambrosetti and Rabinowitz's approach to variational problems, see for example [**AR**], centered around notions like "the deformation lemma"—using different engineered flows to study a given variational problem—and "min-max characterizations". Whereas the Conley approach looked at bounded flow lines and only needed a neighborhood on which one could consider a (local) flow—the focus of consideration is more localized and creates a better intellectual starting point for the endeavor ahead.

All this created cultural barriers which needed to be torn down, or at least ignored, in order to reach the next level.

What Floer took away from his work with the Conley index theory was that a compact space of bounded flow lines is important and carries crucial information if the space can be isolated—in the sense that there exists a small open box containing the compact space of bounded flow lines, but no other bounded flow lines. The latter is a technical requirement for Conley's theory, stipulating at least a local flow picture beyond the bounded flow lines. Conley used this local structure for the construction of the Conley index. For Floer theory, one needs to get away from the isolation requirement; meaning that we only use bounded flow lines but not "their surroundings"—an upcoming cultural shift. This is illustrated in Figure 1.

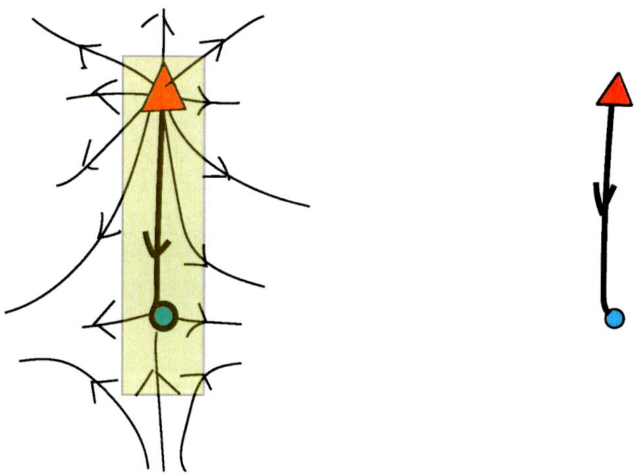

FIGURE 1. On the left is a local flow; the isolated closed invariant set consists of two rest points (constant flow lines) and a flow line connecting them. For the construction of his index, Conley used what is contained in the yellow rectangle, with the index independent of the rectangle as long as it satisfies some properties. Later on, Floer used only what is shown on the right: two rest points and the connecting orbit. In the background, however, there is nontrivial overhead: a non-degenerate Fredholm problem. Whereas on the left one needs the local flow, on the right it does not matter where the data comes from—it can be obtained from a partial differential equation, a situation which might not produce the additional data seen on the left. The Witten complex does this already in the finite-dimensional case; and Floer showed later in [**F3**] that these two approaches produce the same information on the homology level.

Floer's Shift. Floer theory instigated a shift which is captured by our slogan "bye-bye to flows:"

$$\text{FLOW} \Longrightarrow \text{NO FLOW}.$$

But this considerably undersells the story. Floer's ideas opened up the study of new mathematical objects going far beyond what one could have imagined in the variational world.

How did Floer get there? He brought together a culture clash of ideas, both on a philosophical as well as technical level.

9.2. Visiting Zehnder's Head First

First, let's recreate Zehnder's thinking as he tried to extend the torus case. Hindsight provides unnatural clarity, so in attempting to get a sense of the real picture at the time, let's add some smoke grenades.

The starting point is again the torus case and the study of a Hamiltonian system associated to the smooth Hamiltonian

$$H : S^1 \times \mathbb{R}^{2n} \to \mathbb{R},$$

where we assume that H is $\mathbb{Z}^{2n}$-periodic in the $\mathbb{R}^{2n}$-coordinates

$$H(t, z + k) = H(t, z) \quad \text{for all } z = ((q_1, p_1), \ldots, (q_n, p_n)) \in \mathbb{R}^{2n} \quad \text{and} \quad k \in \mathbb{Z}^{2n}.$$

In this case we need to find critical points of the map

$$\Phi_H : C^\infty(S^1, \mathbb{R}^{2n}) \to \mathbb{R} : z \to \int_{S^1} \left[\frac{1}{2} \langle -i\dot{z}(t), z(t) \rangle - H(t, z(t)) \right] \cdot dt. \tag{26}$$

We can also write Φ_H more explicitly also as

$$\Phi_H(z) = \frac{1}{2} \cdot \int_{S^1} \sum_{j=1}^{n} [q_j(t)\dot{p}_j(t) - p_j(t)\dot{q}_{j+n}(t)]dt - \int_{S^1} H(t, z(t))dt.$$

In the following we sometimes suppress the dependence of Φ_H on H and just write Φ.

9.2.1. Towards the Unregularized Gradient Flow.

Conley and Zehnder, in their approach, carry out a finite-dimensional reduction in order to study its finite-dimensional gradient flow. We shall denote by z a loop in $C^\infty(S^1, \mathbb{C}^n)$ and shall identify the $2n$-dimensional vector space $\mathbb{R}^{2n}$ with the n-dimensional complex vector space $\mathbb{C}^n$ (viewed as a real $2n$-dimensional vector space) via

$$(q_1, p_1, \ldots, q_n, p_n) \equiv (q_1 + ip_1, \ldots, q_n + ip_n) \equiv z.$$

We can present their combined procedure—carrying out a finite-dimensional reduction and then considering its gradient flow—as a system of equations

$$\begin{aligned} 0 &= -i\dot{z}_K^- - P_K^- H'(t, z) \\ 0 &= -i\dot{z}_K^+ - P_K^+ H'(t, z) \\ \frac{\partial z_K^0}{\partial s} &= -i\dot{z}_K^0 - P_K^0 H'(t, z). \end{aligned} \tag{27}$$

Here $i = \sqrt{-1}$ is the most famous complex number, fitting into Euler's beautiful formula combining the "magnificent five" 0, 1, π, e and i :

$$e^{i \cdot \pi} + 1 = 0.$$

Further $H'(t,z)$ denotes the $\mathbb{R}^{2n}$-gradient of H with respect to the spatial variables. Moreover $P_K^{\pm}$ and P_K^0 are orthogonal projections acting on a loop $z \in C^\infty(S^1, \mathbb{C}^n)$, written as $z = \sum a_k \cdot e^{2\pi ikt}$, as follows

$$\begin{aligned}(28)\quad z = z_K^- + z_K^0 + z_K^+ &\equiv P_K^-(z) + P_K^0(z) + P_K^+(z)\\ &= \sum_{k<-K} a_k \cdot e^{2\pi ikt} + \sum_{-K\le k\le K} a_k \cdot e^{2\pi ikt} + \sum_{k>K} a_k \cdot e^{2\pi ikt}.\end{aligned}$$

The meaning of $\dot z$ is that we take the derivative with respect to t. Given $z = \sum a_k \cdot e^{2\pi ikt}$ we see that

$$\dot z = 2\pi i \cdot \sum_k a_k \cdot k \cdot e^{2\pi ikt} \quad \text{or equivalently} \quad -i\dot z = 2\pi \cdot \sum_k a_k \cdot k \cdot e^{2\pi ikt}.$$

The equations in (27) constitute an odd system: Writing $z = P_K^- z + P_K^0 z + P_K^+ z$ we obtain some kind of flow $s \to z(s)$, where $s \in \mathbb{R}$ and $z(s)$ for every s is a loop in $\mathbb{R}^{2n} \equiv \mathbb{C}^n$. Hence z can be viewed as a map

$$\mathbb{R} \times S^1 \to \mathbb{C}^n : (s,t) \to z(s)(t).$$

This means, in particular, that z can be written

$$z(s,t) = \sum_k a_k(s) \cdot e^{2\pi ikt}.$$

We note that

$$(29)\qquad P_K^0 \frac{\partial z}{\partial s} = \frac{\partial z_K^0}{\partial s}$$

and therefore the system in (27) satisfies (30):

$$(30)\qquad P_K^0\left(\frac{\partial z}{\partial s}\right) + i\frac{\partial z}{\partial t} + H'(t,z) = 0.$$

It is interesting to note that if we formally (without worrying whether it makes mathematical sense) take the limit $K \to \infty$ we obtain the equation $\frac{\partial z}{\partial s} + J\frac{\partial z}{\partial t} + H'(t,z) = 0$ which we can rewrite as

$$(31)\qquad \bar\partial(z) + H'(t,z) = 0.$$

Here $\bar\partial = \frac{\partial}{\partial s} + i\frac{\partial}{\partial t}$ is the standard linear Cauchy-Riemann operator—a fundamental differential operator in the theory of complex analysis.

This partial differential equation (31) is precisely the Cauchy-Riemann equation perturbed by the Hamiltonian which breaks the symmetry and which upset Gromov at the Il Ciocco conference, see the Prologue on page xvi.

While trying unsuccessfully to locally implant ideas used in the torus case into a more general situation, it crossed Zehnder's mind to replace (27) by taking the previously mentioned formal limit $K \to \infty$ and to look for bounded flow lines.

This means to study the perturbed Cauchy-Riemann equation:

$$(32)\qquad \frac{\partial u}{\partial s} + i\frac{\partial u}{\partial t} + H'(t,u) = 0,$$

where u denotes a map $\mathbb{R} \times S^1 \to \mathbb{C}^n$ or $\mathbb{R} \times S^1 \to T^{2n}$ and $s \to \Phi_H(u(s,\cdot))$ should stay bounded. There are other possibilities for what one may view as a bounded solution, but any reasonable notion turns out to be equivalent.

For anybody in the small group of mathematicians initially involved in the Arnold conjecture, this is a new type of equation. Even if it involves the classical

Cauchy-Riemann operator, it hadn't been studied in these circles. Studying this equation would be labor-intensive and the outcome—would it be worth the time?—was uncertain.

Also, the study of $\bar{\partial}$ had mostly been done by a different group of people in complex analysis with a quite different focus. What was needed here had not been studied before. The closest starting point was the linear theory for generalized analytic functions, developed in the (almost) unreadable book by Il'ya Nestorovich Vekua, [**Ve**][113] —this could have at least provided the local theory, albeit a far way from what was needed.

In the flow-type approaches, it is not too difficult to establish the existence of global flow-lines as a consequence of the well-developed theory of ordinary differential equations, [**Am1**]. The advantage here is that we at least have the existence of the mathematical object—at that time, this was not the case for the equation displayed in (32).

9.2.2. Bounded Flow Lines. Not worrying about the technical details of existence, assume that we have a global solution of (32). The local theory of Vekua would imply that u is a smooth map. As we know from Conley's index theory, we should be interested in bounded solutions—but what is a suitable definition for a bounded solution? After some thought, one can come up with a candidate appealing to the variational structure associated to (32).

DEFINITION 9.2.1. *A bounded flow line for the equation* (32) *is a smooth map*

$$u : \mathbb{R} \times S^1 \to \mathbb{C}^n$$

such that

$$-\infty < \inf_{s\in\mathbb{R}} \Phi(u(s)) \leq \sup_{s\in\mathbb{R}} \Phi(u(s)) < +\infty. \tag{33}$$

Assuming this condition holds, and denoting suitable upper and lower bounds on $s \to \Phi(u(s))$ by $B^{\pm}$, we obtain the following estimate:

$$\begin{aligned} B^+ - B^- &\geq \int_{-\infty}^{+\infty} \left(\frac{d}{ds}\Phi(u(s))\right) \cdot ds \\ &= \int_{-\infty}^{+\infty} \left\langle \nabla_{L^2}\Phi(u(s)), \frac{d}{ds}u(s)\right\rangle_{L^2} \cdot ds \\ &= \int_{-\infty}^{+\infty} \|\nabla_{L^2}\Phi(u(s))\|^2_{L^2(S^1,\mathbb{R}^{2n})} \cdot ds \\ &= \int_{\mathbb{R}\times S^1} \left| -i\frac{\partial u}{\partial t}(s,t) - \nabla H_t(u(s,t))\right|^2_{\mathbb{R}^{2n}} \cdot dsdt \\ &= \frac{1}{2}\cdot \int_{\mathbb{R}\times S^1} \left[\left|\frac{\partial u}{\partial s}(s,t)\right|^2 + \left|-i\frac{\partial u}{\partial t}(s,t) - \nabla H_t(u(s,t))\right|^2\right] \cdot dsdt. \end{aligned} \tag{34}$$

In Floer's hands, this becomes a crucial estimate. From an analysis viewpoint, it is a border line estimate, which does not exclude bubbling-off (à la Uhlenbeck) but allows one to control it in certain cases through topological assumptions. In the torus case one can show that bubbling-off does not occur.

However, nobody, at this point, has a method to find such u.[114]

* * *

In short, in 1983 Zehnder seeded the following idea with Floer: Directly investigate the bounded solutions of (32).

But nobody knew whether that was a *good* idea or not. And even if *good* just means we can sometimes construct a nontrivial bounded flow line, this clearly would have required a lot of work. Furthermore, proving a result like the Conley-Zehnder theorem with such methods would also be a lot of work.

Although Floer does not investigate this question at first, he was interested in the Arnold conjecture and began studying the surface case using variations of the Conley-Zehnder approach. As a by-product, he understood the Conley-index theory in a short time, and published a paper on the subject, [**F6**].

In hindsight, by early 1983, it was possible with the existing mathematical tools to prove the existence of bounded unregularized flow lines in the Hamiltonian context for T^{2n}.[115] However, the conviction wasn't there and so it did not happen. The mathematical world had to wait until 1985—a relatively short interval compared with paper refereeing timelines—for Gromov to come up with his pseudoholomorphic curve theory. This idea triggered Floer's breakthrough on the unregularized gradient flow.

9.3. Arrivederci, Flows!

When Zehnder planted the seed about bounded flow lines for the unregularized gradient flow, Floer had already learned from the Donaldson seminar about some of the analytical techniques that would be crucial for the development of Floer theory. A topologist by training, the seminar forced Floer to learn the theory of elliptic partial differential equations and more advanced ideas, such as Uhlenbeck's bubbling-off analysis and Taubes' gluing analysis.

As far as his knowledge of Witten's Morse theory is concerned, we can only speculate. Witten's paper was talked about in 1982 and 1983—perhaps Floer read it just out of curiosity around this time. It is also possible that Floer became familiar with the idea that homology can be computed from the combinatorics of rigid flow lines through Milnor's book on the h-cobordism theorem, [**Mi2**], and only later learned of Witten's paper through Atiyah, at the Arbeitstagung in June of 1986. In any case Floer, while at Courant, told Viterbo that Witten's paper was important for the development of his theory.

Of the papers [**F0, F1, F2, F3, F4**] that constitute Floer's work on the Morse version of the Arnold conjecture, his paper on the Witten complex, [**F3**], is the last to be submitted, in January 1988.[116] Floer's papers do not make clear when he learned that the combinatorics of bounded flow lines determines homology (in the generic case).[117] This is certain: In January 1985, when Gromov's pseudo-holomorphic preprint, [**G**], arrived, Floer knew that the homology of a compact smooth manifold could be computed in terms of the rigid flow lines associated to a Morse-Smale flow.

Gromov's paper was quickly digested in Weinstein's seminar. Gromov had shown that by studying

$$\bar{\partial}_J u = 0, \tag{35}$$

a generalization of $\bar{\partial} u = 0$ can be done in very large generality, and it can be used to address long-standing symplectic questions.

At this point Zehnder's suggestion (page 87) would have started making sense to Floer—since Gromov's success indicated that there might be a theory for finding bounded gradient lines for the unregularized gradient flow.

Zehnder's idea of studying

$$\bar{\partial}u + H'(t,u) = 0 \quad \text{and} \quad -\infty < \inf_{s\in\mathbb{R}} \Phi(u(s)) \leq \sup_{s\in\mathbb{R}} \Phi(u(s)) < +\infty \tag{36}$$

initially for the torus suffered from the dire fact that it wasn't even clear how to approach this without substantial work and what the outcome of the work would be was not clear at all—any undertaking in this direction looked very risky. Whereas Gromov had shown that an equation of the form

$$\bar{\partial}_J(u) = 0$$

could be studied on any (reasonable) symplectic manifold, among them the compact ones—although there were still many problems when bubbling-off was involved.

With Floer having acquired analytical skills, he was convinced that one could study equations of the more general type

$$\bar{\partial}_J u + H'(t,u) = 0 \tag{37}$$

on quite general symplectic manifolds. This type of equations are now called "Floer equations." Even if the evolution towards that equation looks natural, it turns out such an extension is not easy. And of course, Gromov, upon hearing of Floer's ideas, thought they were complete nonsense. Gromov's paper also made clear that a bubbling-off analysis à la Uhlenbeck was an intrinsic part of studying pseudoholomorphic curves, and clearly this is going to show up in the study of (37) as well. Gromov's work also showed that suitable topological assumptions on the underlying symplectic manifold would prevent bubbles from forming, simplifying the analysis.

Floer concluded that on many symplectic manifolds a bubbling-off analysis for (37) could be carried out and that tools could be developed to find bounded flow lines for the unregularized gradient flow. He convinced himself that the ends of the bounded flow lines converge in the non-degenerate case to periodic orbits.

He reckoned that with some luck the equation in (37) will be of Fredholm type analogous to the Flow equation in Morse theory, where the Fredholm index is the difference of the Morse indices. Conley and Zehnder had explained in previous work that the "difference of the (infinite) Morse indices" is the difference of the Conley-Zehnder indices—so the Fredholm index of (37) at a solution u connecting two periodic orbits of Conley-Zehnder indices a and b should be $b-a$.

During the spring of 1985, the pieces of the puzzle came together. Floer had a precise plan of attack on the problem. This plan assumed a favorable outcome for some of the technical points that yet needed resolution—ultimately realized decisively during the 1986-1987 academic year when Floer was an instructor at the Courant Institute. Even in June 1986, things looked promising as Floer delivered his talk on the Arnold conjecture at the prestigious Arbeitstagung, see page 83. And then that August the Bulletin of the AMS received the announcement of his results about the Arnold conjecture, [**F0**]. This three-page paper did not provide too much detail; Floer made no mention of the Witten paper, [**Wi**], but he refers to Milnor's book on the h-cobordism theorem, [**Mi2**], for the part of the argument about counting only rigid flow lines. Maybe this is an instance of sloppy citations, or maybe it is an indication of the sources Floer knew to date.

It was June 1987 before Floer submitted his technical paper on the unregularized gradient flow which lies at the heart of all of his constructions. On the odd technical question, Floer had consulted Courant analysts—he was known to get flak for some of his idiosyncratic self-taught knowledge. Louis Nirenberg, who later won an Abel Prize, was no doubt a valuable resource. Another useful point of reference was a paper by Lockhardt and McOwen [**LoMcO**]. Concerned with the Fredholm property of linear elliptic partial differential operators with asymptotic boundary conditions, it provided one of the inputs he needed: the equations for the bounded unregularized gradient flow are Fredholm and, indeed, the Fredholm index is $b - a$.

Floer's first results from his investigations of the linear theory and the computation of the Fredholm index in terms of asymptotic Conley-Zehnder indices were published in [**F4**]—one of the four core papers establishing the Arnold conjecture in the non-degenerate case under some topological assumption on the symplectic manifold. The second paper in the series, [**F1**], was the treatment of the unregularized gradient flow. It resolved the analytical issues—[**F4**] is an input to [**F1**], providing the computation of the Fredholm index and (virtual) dimension of the space of bounded orbits from x to y in terms of the Conley-Zehnder indices. The third paper,[**F2**], the pièce de résistance, delivered the spectacular result: The Arnold conjecture for a quite general class of Lagrangian intersection problems. But it was contingent upon his final paper on the Witten complex,[**F3**] submitted in January 1988.

In the introduction of the Witten paper he described in detail the historical account and attributed the idea of counting rigid flow lines to Milnor, [**Mi2**], but also noted that it was known to Smale and Thom. He mentioned that it did not receive a lot of attention, and that Witten, [**Wi**] rediscovered it, giving a physicist's proof made precise by Helffer and Sjöstrand in [**HeS**]. He called it the Witten complex due to [**Wi**]. In [**F3**] he gave a proof of the fact that one only needs to count rigid flow lines by deriving the properties of the Witten complex using the Conley index theory. He pointed out the relationship to the concept of a connection matrix and Robert Franzosa's work [**Fran**] (writing consistently "Fransoza" instead of "Franzosa"; Franzosa, Ph.D. 1984, is one of the last three graduate students of Conley).).[118]

In total, it took Floer several papers, written over a period of three years, 1985 to 1988, to square away all the vulnerable technicalities for his proof of the Arnold conjecture for a more general class of compact symplectic manifolds. So, for quite a long time, Floer himself had to hope and believe that certain things would work out.

In his breakthrough work Floer had proved, in a situation where one is able to exclude bubbling-off by topological reasons, that one can use the critical points and solutions of a perturbed Cauchy-Riemann equation to define a directed decorated graph to which the algebraic procedure of taking homology can be applied. He also is able to compute the homology. The amount of high-level analytical work he has to perform is staggering. In the appendix on page 147 we discuss in more detail some of the mathematical issues Floer had to overcome.

CHAPTER 10

Symplectic Lift-Off

As Floer's ideas continued to evolve, new opportunities arose. Some long-standing problems, previously deemed inaccessible, could now be studied. Floer's appealing ideas could also be transported into new contexts; and linked to adjacent notions in clever interlocking ways. For instance, a fine development—motivated by Gromov's non-squeezing result and further work by Ekeland and Hofer—came together in the Jardin du Luxembourg.

10.1. The Jungle Invades the Jardin

Gromov had proved, in his 1985 paper [**G**], his famous non-squeezing result, as an application of his pseudoholomorphic curve theory. Namely, denote by $B^{2n}(r)$ the open Euclidean ball of radius r in the standard symplectic vector space $(\mathbb{R}^{2n}, \omega_0)$ with coordinates $z = (z_1, \ldots, z_n) = ((q_1, p_1), \ldots, (q_n, p_n))$, i.e.

$$B^{2n}(r) = \{z \in \mathbb{R}^{2n} \mid |z| < r\}$$

and define the symplectic cylinder $Z^{2n}(R)$ by

$$Z^{2n}(R) = \{z \in \mathbb{R}^{2n} \mid |z_1| < R\}.$$

Gromov proved the following remarkable theorem:

THEOREM 10.1.1 (Gromov). *If there exists a symplectic embedding* $\phi : B^{2n}(r) \to Z^{2n}(R)$, *then* $r \leq R$.

This theorem answers two questions (Question I and Question II on page 191 in [**Fe**]), raised in 1983 by Charles Fefferman in the context of his 'SAK theory' estimating the number of eigenvalues for differential operators. We recall these questions below and mention that with the progress made on embedding obstructions it might be worthwhile to revisit Fefferman's work. Possibly a novel mixture of metric and symplectic geometry will be useful—it might be worth a try!

The Fefferman questions:

> *Question* I. Can the unit cube in $\mathbb{R}^4$ be embedded by a canonical transformation into $T_\delta = \{(x, y, \xi, \eta) \mid |x|, |\xi| < \delta,\ y, \eta \text{ arbitrary}\}$? Here $\delta << 1$.
> *Question* II. Can T_δ above be embedded by a canonical transformation into $T_{\delta'}$ with $\delta' < \delta$.

As a consequence of the non-squeezing result, the answer to both questions is "No." Moreover, Gromov's result implied that in dimension at least 4 one can always find a volume preserving diffeomorphism that cannot be approximated in the C^0-topology by a symplectic diffeomorphism—a very striking result.

Gromov's result prompted research by Ekeland and Hofer, bringing Gromov's theorem into a larger context. In the fall of 1987, Ekeland visited Hofer for a month at Rutgers; they commuted daily to Princeton, since Hofer was a member at the

Institute for Advanced Study.[119] Ekeland and Hofer, prior to Gromov's work, had studied convex Hamiltonian systems, [**EH4**], which led to investigations, [**EH3**], about the relationships of local invariants of periodic orbits on prescribed energy surfaces and global invariants which can be associated to the energy surface. With the context given by Gromov, these papers triggered a powerful new idea, namely that one should be able to use Hamiltonian dynamics to construct obstructions for symplectic embeddings. During this visit they introduced the notion of a "symplectic capacity", and they constructed what are now called the "Ekeland-Hofer capacities"[**EH1**, **EH2**]. Later this work was followed by the construction of the Hofer-Zehnder capacity in [**HZ3**]. The new understanding resulting from this work formed the basis for [**H4**] which introduces what is now referred to as Hofer geometry. Several years later important work by Lalonde and McDuff, [**LMc1**, **LMc4**], placed this geometry, among other things, in the context of a generalized non-squeezing result—a non-obvious connection.

Next, we describe some of the details concerning symplectic capacity theory. Restricting ourselves to the symplectic vector space $(\mathbb{R}^{2n}, \omega_0)$ we are interested in introducing certain measurements that quantify what one might call the symplectic size of open subsets of $\mathbb{R}^{2n}$. For the following, we denote by $\mathcal{O}^{2n}$ the collection of all open subsets of $\mathbb{R}^{2n}$.

DEFINITION 10.1.1. *A symplectic capacity is a map* $c : \mathcal{O}^{2n} \to [0, +\infty]$ *having the following properties:*

(1) $0 < c(B^{2n}(1)) \leq c(Z^{2n}(1)) < \infty$ *(nontriviality).*
(2) *If there exists a symplectic diffeomorphism* $\Phi : \mathbb{R}^{2n} \to \mathbb{R}^{2n}$ *satisfying* $\Phi(U) \subset V$ *for* U, $V \in \mathcal{O}^{2n}$ *then* $c(U) \leq c(V)$ *(monotonicity).*
(3) $c(\alpha \cdot U) = \alpha^2 \cdot c(U)$, *where* $\alpha \in \mathbb{R} \setminus \{0\}$ *and* $U \in \mathcal{O}^{2n}$ *(conformality).*

The existence of a capacity satisfying $c(B^{2n}(1)) = c(Z^{2n}(1))$ follows from Gromov's non-squeezing theorem. Ekeland and Hofer showed that there exists a sequence of symplectic capacities

$$c_1 \leq c_2 \leq \dots$$

with these properties: Namely,

$$c_k(Z^{2n}(1)) = k \cdot \pi \quad \text{and} \quad c_k(B^{2n}(1)) = (1 + [(k-1)/n]) \cdot \pi,$$

where $[p]$ denotes the integer part of a real number $p \geq 0$. The sequence also has the following property: Starting with an open ellipsoid,

$$E(a) = E(a_1, a_2, \dots, a_n) = \left\{ z \in \mathbb{R}^{2n} \;\middle|\; \sum_{i=1}^{n} \frac{q_i^2 + p_i^2}{a_i} < 1 \right\},$$

where $a = (a_1, \dots, a_n)$, with $0 < a_1 \leq \dots \leq a_n < \infty$, order the set of numbers $\{j \cdot a_i \mid j = 1, 2 \dots$ and $i = 1, \dots, n\}$ in non-decreasing order, repeating elements according to their multiplicity, i.e. counting the number of ways they can be written as $j \cdot a_i$. This gives a sequence $\ell_k(a)$, $k = 1, 2, \dots$ and clearly $\ell_1(a) = a_1$.

For example, if $a = (1, 2, 2)$, one obtains the sequence $\ell_k(1, 2, 2)$:

$$1, 2, 2, 2, 3, 4, 4, 4, 5, 6, 6, 6, \dots$$

In fact, it was shown that the capacity sequence also satisfies the identity

$$c_k(E(a)) = \pi \cdot \ell_k(a).$$

Moreover, the construction of the c_k implied another property. Namely, given a bounded connected open subset $U \in \mathcal{O}^{2n}$ with a smooth boundary ∂U, take a Hamiltonian $H : \mathbb{R}^{2n} \to [0, \infty)$ with the following features:

(1) $H(z) = 1$ and $dH(z) \neq 0$ for $z \in \partial U$
(2) $H(z) < 1$ for $z \in U$
(3) $H(z) > 1$ otherwise

Next one considers the periodic orbits of the Hamiltonian vector field X_H with energy $H = 1$. More precisely, consider pairs (z, T) where $z : \mathbb{R} \to \mathbb{R}^{2n}$ satisfies

$$\dot{z} = X_H(z) \quad \text{and} \quad H(z(t)) = 1, \tag{38}$$

and further $T > 0$ such that $z(0) = z(T)$. The "action spectrum," denoted by $\sigma(U)$, is then defined as the set of all numbers $\int_0^T \lambda(X_H(z(t))) \cdot dt$, $d\lambda = \omega$, coming from the pairs (z, T).[120] It can be shown that $\sigma(U)$ does not depend on the choice of H as long as U is a bounded connected open subset of $\mathbb{R}^{2n}$ with a smooth boundary.

We say that U is of "restricted contact type" provided there exists a one-form λ satisfying $d\lambda = \omega$ so that $\lambda(X_H(z)) > 0$ for all $z \in \partial U$. Again, this property does not depend on the choice of H as long as it has the properties previously described. The constructions of the c_k by Ekeland and Hofer give the additional property that

(4) $c_k(U) \in \sigma(U)$ for all $k = 1, 2, \ldots$ provided U is of restricted contact type.

In other words the numbers $c_k(U)$ are actions of periodic orbits on the energy surface ∂U and the construction connects symplectic rigidity questions with dynamical features on an energy surface in view of the axioms satisfied by the c_k.

The construction also has some continuity features. Given U of restricted contact type and $k \geq 1$ one can find for given $\varepsilon > 0$ open V^- and V^+ of restricted contact type such that

$$\mathrm{cl}(V^-) \subset U \subset \mathrm{cl}(U) \subset V^+,$$

and $c_k(U) - \varepsilon < c_k(V^-)$ and $c_k(V^+) < c_k(U) + \varepsilon$. This implies a certain continuity property of the maps $U \to c_k(U)$ on the subset of contact type domains provided the set of such domains is equipped with a suitable topology.

By early 1988, Floer had lightened his burden—having finally submitted his flight of papers proving the Arnold conjecture under a topological assumption which rules out bubbling-off; precisely the result he'd originally claimed. Now he was thinking about removing the topological assumptions in the symplectic case, as well as the version of Floer theory for the Chern-Simons functional, as suggested by Atiyah—and he was again open to other mathematical "distractions."

While at Courant that spring, Hofer told Floer about the work he had done with Ekeland. Together they started thinking about how the symplectic capacities could be obtained from a version of Floer homology. Fortuitously, Ekeland had invited Floer and Hofer to be visiting professors at the University Paris Dauphine for June 1988: The weather was excellent, and so was their progress. Floer and Hofer spent their days in the Jardin du Luxembourg discussing the relationship between the two theories. It soon became clear that there was a relationship to an S^1-equivariant version of Floer homology for symplectic domains.[121] The collaboration was very fruitful and the results became part of the opening salvo of the symplectic program held at the MSRI in the fall of 1988. (Floer gave a brief account of this collaboration in the Berkeley Summer/Fall Progress report reproduced on page 123.)

10.2. "Capacities, etc"

As his fame grew, Floer had had a stressful if productive time at Courant. His appointment there took him through spring of 1988. However, before he'd even addressed himself to the task of applying for another job, he was courted, and received three offers. The first proactive action came from Rutgers University: an offer of full professorship with tenure. This prompted a counter-offer from Courant: an associate professorship with tenure. Knowing what was on the table, Berkeley "strikes back hard:" assistant professor, tenure-track.

He started in the fall of 1988 at Berkeley—with the low-end offer he reasoned that expectations couldn't be too high; he'd be able to work without the distractions of a higher position's demands. And he would be reunited with some of his friends from Barrington Hall. Although, one of yesteryear's police raids on Barrington came back to haunt him: Since he had been arrested, he encountered problems when applying for a green card; he hired a lawyer and the matter was somehow resolved.[122]

Berkeley's seemingly stingy and arrogant offer had turned out to be a stroke of genius since Floer didn't care about money. He gladly gave it away—to friends in need, or simply in want (he bought one of his pals a piano). Floer even bailed Hofer out when he arrived in Berkeley for the MSRI symplectic program and found himself short of money. In those days, the ATM systems were still in the works. The institute would pay a per diem after registration (and an audience with 'Kap,' Irving Kaplansky, then the director). Checks would be produced on a regular basis, which could be turned into greenbacks in a specified bank downtown. Unfortunately, with a weekend in between, Hofer foresaw a cash flow problem. Fortunately, he'd arranged to meet with Floer upon arrival, and he asked if his friend could lend him some money. With a serious face, Andreas reached into one of his trouser pockets and handed over a wad of bills: "Ist das genug?"—Was it enough? The amount, in excess of $1000, seemed plenty. When asked if he made a habit of running around town with such large sums of money, Andreas explained that one of his friends had decided to turn his life around and start a company—that day Andreas had intended to give him some startup money, but his friend was a no-show for their meeting. Andreas was also known to give generously to homeless people. He was happy to give away almost all of his money.[123]

The symplectic program ran from August 1988 through July 1989. Organized by Robert Devaney, Hermann Flaschka, and co-chairs Victor Guillemin and Alan Weinstein, the gathering embodied a breadth of sensibilities.[124] A new domain was in the making—symplectic topology—and the wide ranging perspectives were a tremendous asset. The proceedings began with a Hofer-Floer four-hour tag-team talk: HFHF. Hofer expounded on "Symplectic Capacities" and Floer followed with "Capacities, etc." Hofer described his work with Ekeland—the axioms for a capacity and the construction of infinitely many such capacities. Floer then described how one could obtain these results from introducing an S^1-equivariant version of Floer theory, laying the groundwork for an equivariant symplectic homology theory (never written up).[125] Hofer described in a later talk the construction of symplectic homology and its properties.[126] Some of the statements in Floer's talks were later put on a solid footing as a special case of results in a paper by Jean Gutt and Michael Hutchings, [**GuH**].

And with that we had lift-off: symplectic capacities and symplectic homology became indispensable tools in symplectic geometry. A more general version of S^1-equivariant symplectic homology is later outlined by Viterbo in [**Vi4**]. Symplectic homology got a great boost by Paul Seidel, see [**Sei, Per**], who popularized its use in homological mirror symmetry. Bourgeois and Oancea define S^1-equivariant symplectic homology in [**BouO1**] and later in [**BouO2**] relate it to a version of contact homology which is part of the SFT-family of Floer-type theories introduced in [**EGH**]. In [**HutN1**] Hutchings and Nelson describe a general scheme for constructing S^1-Morse-Bott-Floer theories and apply the ideas in the construction given in [**HutN2**].[127]

10.3. Back In Bochum

In the summer of 1989, Hofer moved to Germany to start a professorship at Bochum. Jointly with the complex analyst Alan Huckleberry,[128] he organized a big international conference in the first half of 1990, within the "Sonderforschungsbereich" (a type of research institute) between theoretical physicists and mathematicians. Huckleberry was in charge of publicity for the event. The headlines in the Ruhrgebiet's newspapers announced, essentially: "Superhirne kommen nach Bochum"—"Superbrains come to Bochum." All the usual suspects were in attendance, including Floer.[129] One day, Floer received a letter from Göttingen inviting him to apply for a vacant position. Viterbo and Hofer having lunch with him in the cafeteria when he opened the letter. Memorably, he left it with the dirty dishes on the conveyor belt. Viterbo commented: "The chef will be happy receiving an offer from Göttingen." Floer did not apply and quite likely never answered the letter. But in the moment, this raised the possibility that he might move to Bochum. Somewhat unexpectedly, he showed great interest—even though such a position would be a full(!) professorship. Rolf Kaiser speculated that Floer might have been keen on such a move since it would get him closer to his friend Hélène McGiffin of his Barrington Hall days, who had since moved to Paris.

One of the problems with hiring Floer at Bochum was that there was no available professorship and the German system at that time made opportunistic hirings difficult. Nonetheless, Huckleberry, then the dean, was easily convinced that the department should try to create a position for Floer.[130] And sure enough, Huckleberry, together with the rector of the university, negotiated with the Ministry of Science and Research and an additional position was made available in the record time of about six weeks.[131]

Following the usual protocol, the position was advertised, but of course Floer was offered the job; he was light years ahead of any other applicant.[132] He was then invited to meet with the powers that be and discuss the matter of money; the details of research funding with the chancellor and his salary with a Staatssekretär (secretary of state) from the ministry. He arrived the day before and stayed with the Hofer family—Hofer's wife Sonja and their kids knew Andreas from his Rutgers visits. After a nice dinner Hofer coached Floer on the negotiation: Explaining concepts like Vordienstzeiten, Dienstjahre, Zuschläge, Sonderzuschläge, Pensionsfähigkeit von Bezügen and Kaufkraftparität. In short, the main message was that even if he didn't care about the money, this was the point where he had considerable leverage; and that of course he would be free to give all his money away. Anticipating that his negotiating skills might not be stellar, the ministry had been informed through unofficial channels that they should do everything possible to bring this young genius back to Germany.[133]

The morning of the appointment, Andreas came down from the bedroom and instead of his usual T-shirt he was wearing a nice outfit—Hofer had advised that proper attire would be a good idea, and he had seriously dressed up. There was one item of concern that he asked Sonja about: Floer explained that he shaved everyday, but in the evening, so as to gain valuable minutes in the morning when he felt in top form intellectually. He solicited Sonja's opinion about whether he should shave before leaving for this occasion. After Sonja's response—"Andreas, unless you are going to kiss the Staatssekretär, you should be fine"—he laughed and felt at ease. Hofer brought him to the station in Recklinghausen to catch a train to Düsseldorf, the capital of Nordrhein-Westfalen. When he picked him up toward the end of the day Andreas was clearly pleased. He started in Bochum in the summer of 1990.

That August, the International Congress of Mathematicians convened in Kyoto. Andreas had been invited to give a plenary lecture—given the rumors that he'd been discussed for the Fields Medal, this was a decent sign that he might be in the running, for 1994 if not for 1990. When Floer visited Zehnder in Zürich before flying to Japan, he did a test run for his lecture. The first attempt ran much longer than the allotted hour, whereas the second one was too short. Criticizing both attempts, Zehnder hoped for the best. Once at the ICM, Floer's lecture went perfectly, addressing: "Elliptic methods in variational problems", [**F12**].

Afterwards, the Hofers invited Floer to join them for some sightseeing. But outside the Congress itinerary he was quite busy. They later learned that Floer's friend Hélène had joined him in Kyoto; he had never once mentioned Hélène, let alone that she often accompanied him on his international travels. He had organized his life into parallel universes that did not intersect.[134]

CHAPTER 11

Thereafter

Floer's arrival in the Bochum math department was much anticipated. Kai Cieliebak, then a graduate student, knew of Floer and his work—and Floer had been announced in grand terms. As Cieliebak recalled: "Huckleberry told us, 'Hofer is a great mathematician. But wait until the really great one comes.' In German, he said, 'Gegen Floer ist der Hofer ein Waisenknabe'—'Compared to Floer, Hofer is an orphan'."

Cieliebak and his fellow students were eager to set eyes on the living legend. "And then arrives this tiny, shy, timid person—compared to Hofer, who is an imposing, huge guy." That was Cieliebak's first impression. And it didn't change much when he took Floer's lecture course on gauge theory. Floer spoke quietly, with hesitation. He hugged the blackboard. He barely interacted with his audience. "Honestly," Cieliebak said, "it seemed like he was not very well prepared." Cieliebak later learned that in fact Floer had put significant effort into preparation; that he'd thought a lot about how to convey the complexities of the subject. But Cieliebak didn't understand much from Floer's teaching: "I felt totally lost." He dropped the course and taught himself the material.

FIGURE 1. Kai Cieliebak.

For a time, Floer often visited the Hofers and joined them for dinner. He was happy and at ease with their children—Boris (age 14) and Meret (7) and David (4), all of whom adored Andreas. Invariably when dinner was ready, Floer and Meret and David were nowhere to be found. Hofer would organize a one-man search party; he always found them at a playground in the neighborhood.

In early 1991, however, it was apparent that Andreas was struggling. He had long suffered from depression (as had his mother) and his mental health was deteriorating. He seemed restless. He traveled multiple times to Paris to visit Hélène, and Berlin to visit his brother Detlef. His periods of depression consumed longer durations, and left him unable to concentrate, unable to work. At one dinner Andreas mentioned that he was seeing a neurologist. For Sonja, who had studied at the Institute of Applied Psychology in Zürich, this seemed odd; she would have expected him to be treated by a psychotherapist. On one occasion, he mentioned that he was taking medication to treat his depression. A while later, after he returned from a trip to India, he commented to Hofer that the medication didn't work, and that he had stopped taking it. Looking back, Hofer said, "All alarm bells should have gone off at this point. But unfortunately they didn't."

At the start of May, Floer delivered two talks about his instanton homology in Berlin, the first at Humboldt University and the second at the Technical University. At both he drew enthusiastic audiences.[135] It later came to light that during this visit Floer had inquired about jobs in Berlin—he wanted to be closer to his brother, Detlef.[136]

Floer was also apparently in the running for faculty positions further afield—for one, earlier that year Gromov had written a letter of recommendation to Simon Kochen, department chair at Princeton. It read in full:[137]

> *Floer is a superstar. He invented the Floer cohomology which is the most profound thing we have in non-linear global analysis up-to-date, save for Witten's functional integrals (which is not mathematics yet). Also, I believe, that the field opened by Floer will be bringing crops for at least a couple of decades to come. Grab him, whoever has power to do so.*

Once back from Berlin, Floer ran into Hofer in Bochum. Floer was clearly distressed and in poor shape. Hofer invited him to dinner, but Andreas had excuses about why he couldn't come. This was the last time Hofer saw him.

Andreas Floer committed suicide—he jumped to his death from his apartment building, near the university—early on the morning[138] of May 15th, 1991, a Wednesday. Hofer received a phone call in his office at the math department. He returned home to tell Sonja. Meret was there for lunch. She had been born in the United Kingdom and always lived in an English-speaking country until moving to Germany in spring 1989, at which point she switched completely to the German language and seemed to no longer understand English. Trying to protect his daughter from the tragedy, Hofer told Sonja in English: "Andreas killed himself." Meret became completely pale; she understood. Shortly after, she sat down at the piano and played a piece she had practiced with Andreas—the Flea Waltz for four hands.

11.1. Spheres Collide

Soon after learning of Andreas' death a small delegation visited Floer's mother in Duisburg, to offer their condolences—Uwe Storch, then the dean of the faculty of mathematics, together with Huckleberry, Helmut and Sonja. Some three decades later, Helmut observed that coping with a sudden death—especially a violent death—varies dramatically among individuals, as the Hofer family knows from their own experience, [**TE**]. While Floer's mother was composed, no doubt in shock, his eldest brother Rainer was angry. He noted that Andreas hadn't even been capable of buying a decent used car—indeed, he had recently acquired a used car, and in the winter arrived at the Hofers' frozen to a nearly hypothermic state since the car's heating did not work. Andreas' mother mentioned that she now had to sell his viola da gamba.

FIGURE 2. Andreas at work—on the terrace of his parents' house, where he grew up in Duisburg, circa 1990.

It soon became apparent that the family was not aware of Andreas' success as a mathematician. They knew he was a professor at Bochum, but that apparently was of little importance. Hofer steered the conversation in a different direction: He described how Andreas was a mathematical celebrity and that his contributions had changed the mathematical landscape. This made an impression—and a visible difference especially for Andreas' mother.

At the funeral, the spheres of Floer's life—inhabitants of each world mostly unbeknownst to one another—intersected for the first time: Family. Bochum friends such as Rolf Kaiser and his partner Kurt Heim. Mathematicians. There was also a Barrington Hall contingent: Hélène McGiffin came from Paris; she had traveled far and wide with Andreas, taking care of him. Rebecca Steinberg arrived from Berlin—she and Andreas had continued conversations—over pizza and perogies, walks and music—from Berkeley to Brooklyn to Berlin. Leo Sullivan and Peter Spencer and Tanja Zimmerman flew in from Berkeley. For Hofer, who interacted with Andreas mostly in the mathematical dimension, it immediately became clear how little he knew of his friend.

Kaiser throughout the years had pieced together some connections. He knew of Andreas' frequent trips to Paris and his close ties to Hélène. Andreas had confessed to Rolf that he was in love. By various accounts, it was not long before he died that he asked Hélène to marry him—or perhaps, if not a formal proposal (that wouldn't have been his style, Kaiser reckoned), he proposed that they make a life together. Hélène was taken aback by the offer and declined. Their relationship was close, but Platonic; she and Rebecca had assumed that Andreas was gay and in the closet. After unsuccessful attempts at rehab, Hélène committed suicide in 1994.

Kaiser also knew that Andreas' Paris trips included treasured meetings with Gromov. Andreas raved about "Der Gromov"—"The Gromov." In proper German, such a sentence—"Der Gromov hat ein Lemma bewiesen" versus "Gromov hat ein Lemma bewiesen"—would be considered bad form. For Andreas, with his anti-authoritarian streak, that would have been reason enough to use this construction, and a way to signal deep respect.

Knowing Andreas well, Kaiser observed that his dark sides—the depressions, the drugs—played a fatal role. But that Andreas' brighter sides shape his legacy. Said Kaiser: "I am deeply convinced: Everyone—Everyone!—who knew him feels in some way enriched by him. And, in my opinion, his variant of our characteristic generational conflict is also one of his bright sides, and was one of his strengths. He was one of the most unbourgeois people I have ever met. He was entirely peculiar." In Kaiser's estimation, Floer could be allowed to say, with Goethe: "Ich hab' Mein' Sach' auf Nichts gestellt"—"My trust in nothing now is placed..."

11.2. The Signpost

The week after the funeral, while at the university, Hofer ran into Kaiser, who had been visiting Andreas' office. They gathered with Detlef and talked about Andreas—there was a lot to talk about. Later, Hofer and Kaiser met again and listened to music that Andreas appreciated. Kaiser and Andreas had listened to hours and hours of music together. In total over the years, they had probably spent more time listening to music than talking. He recalled that music could bring Andreas to tears, and that he was especially moved by Schubert's 'Der Wegweiser—The Signpost'[139] whose lyrics encapsulated Andreas' enigmatic nature.

Der Wegweiser

Was vermeid' ich denn die Wege,
Wo die ander'n Wand'rer geh'n,
Suche mir versteckte Stege,
Durch verschneite Felsenhöh'n ?
Habe ja doch nichts begangen,
Daß ich Menschen sollte scheu'n, -
Welch ein törichtes Verlangen
Treibt mich in die Wüstenei'n ?
Weiser stehen auf den Straßen,
Weisen auf die Städte zu.
Und ich wandre sonder Maßen
Ohne Ruh' und suche Ruh'.
Einen Weiser seh' ich stehen
Unverrückt vor meinem Blick;
Eine Straße muß ich gehen,
Die noch keiner ging zurück.

The Signpost[140]

Why do I avoid the highways
Where the other travelers go,
Seeking high, forgotten byways
With their rocks and drifting snow?
Are there reasons for prolonging
Any links with human kind?
Why should foolish fits of longing
Make the wasteland that I find?
All the signposts on the highroads
Only point towards a town;
Without rest, I stalk the byroads–
One desire: to lay me down.
There's a signpost I see standing
To a place I cannot spurn;
It's a road that's all-demanding,
By which no-one can return.

When Hofer discovered that The Signpost was among Andreas' favorites, it brought to mind the following: During the International Congress in Kyoto, the Hofers had managed to entice Andreas to join them for one minor bit of sightseeing. They visited several shrines, among them the Heian Shrine. Sonja took a photo of Andreas standing in front of a signpost that pointed in two directions, both saying, "This Way." Unfortunately, the picture was lost—it was lent to Bochum University upon Andreas' death and never returned. Nonetheless, it captured Andreas and his modus operandi perfectly, and resonated all the more in his absence.

FIGURE 3. The sign post which was in the background of the lost Andreas photo.

11.3. Posthumously

Although under pressure Andreas had gotten the heart of his breakthrough down on paper, he left behind a cache of unpublished work, in various stages of completion.

"When I heard that Floer had committed suicide, I understood immediately that his mental health would have been a major factor in his death," said Karen Uhlenbeck, looking back. When she learned, fairly recently, about his drug use—although hard facts about the circumstances are scant—she noted that the combination of substance abuse and mental imbalance cannot be overemphasized as causes for his suicide. "His difficulties in writing up his mathematics was merely a trigger, just one possible trigger and not the underlying cause," she observed. "I am familiar with struggles with mental health and substance abuse. Approximately 50% of people addicted to drugs or alcohol have mental problems and 50% of people with mental difficulties abuse substances. The inability to write up the mathematics is a result rather than a cause, although there is a temptation by everybody to think it is." She also noted that at this stage, Floer was relatively young and inexperienced, and was therefore unlikely to identify the source of his inability to function. She said, "Only later in life after the struggles with mental health and addiction are at least partly recognized, does the looming catastrophe become evident to both the individual and his associates."

The two papers that he'd announced in the Bulletin of the AMS were later edited by Peter Braam and Ian McAllister and included in The Floer Memorial Volume, published in 1995 [**FMem**]. And in the intervening years, especially since the summer of 1988—collaborating with Hofer in Paris—ideas had accumulated pertaining to the connection between Floer homology and symplectic capacities, which had been reported during the MSRI program. The pair had also done technical work connected with transversality issues, which hadn't been resolved in a satisfying way. And more generally, ongoing discussions between Floer and Hofer had spurred ever more insights about a number of important phenomena.

Hofer completed already existing drafts of this joint work, resulting in a substantial number of posthumous publications.[141] The first paper he finished and submitted was within a month of Floer's death: "Coherent orientations for periodic orbit problems in symplectic geometry." It addressed orientation problems which allow certain Floer homology theories to be defined over the integers rather than just $\mathbb{Z}_2$. This work provided input for a second paper that had also been in preparation, "Symplectic homology I: Open sets in $\mathbf{C}^n$." It introduced the notion of symplectic homology, a symplectic invariant for open sets in the symplectic vector space $\mathbb{R}^{2n}$ emerging from the impetus to relate symplectic capacities and Floer theory.

Eventually following up on the results of that month with Floer in the Jardin, Hofer and Kris Wysocki, then a researcher at Bochum, wrote up the paper "Applications of symplectic homology I", [**FHW**]. One of the applications is a proof of a conjecture by Gromov. A collaboration between Dietmar Salamon and Hofer resulted in the paper, "Transversality in elliptic Morse theory for the symplectic action", [**FHS**]—Salamon had noted that the transversality problems were not adequately addressed in the literature; Floer and Hofer had noted the same.

Later Kai Cieliebak joined the fold to realize, "Symplectic homology II: A general construction" and "Applications of symplectic homology II", [**CFH**, **CFHW**]. Having dropped out of his class, and having never spoken a word with Floer, Cieliebak finds it curious to have joint papers with him. That said, Floer theory is still what Cieliebak is after today—"not exclusively, but it's fundamental," he said.

Cieliebak was, and still is, impressed with Floer's grand vision: The sheer daring; that there should be such thing as Floer homology at all. Having the idea was one thing, but actually making it real required intense, technical labor. "It's quite amazing that he did that work," Cieliebak said. "Because to make Floer homology work, you need a whole lot of analysis, functional analysis. And that's not something that came to him naturally. But still, he just dug into it."

One of the neat, analytical details that he developed was the Floer norm on the space of almost complex structures.[142] "It's a little idea," Cieliebak said—but without it, he said, "you could be stuck for years. That's something that happens to minor human beings like myself all the time. Floer just somehow smartly found this twist and brushed aside a whole lot of difficulty." And there were several of those little ideas, he said, without which Floer might well have become intractably stuck any number of times and never finished the project.

Epilogue

In the fall of 2021, Akshay Venkatesh and Jacob Lurie, both faculty in the School of Mathematics at IAS, convened a Floer learning seminar in Princeton. A critical mass of symplectic geometers had converged at the Institute that year, but this gathering was for the benefit of the uninitiated—participants learned as they lectured and lectured as they learned.[143] Venkatesh, for instance, a well-decorated number theorist, had heard of Floer homology—he had heard the word "Floer" in many contexts, without knowing what it meant. "I wanted to learn what it was," Venkatesh said. The goal of the seminar was simply to study the idea and the philosophy via rapid immersion: "I wanted to learn about the simplest possible case. I wasn't looking to understand the most recent thing, or the most impressive. And so we went back to one of Floer's original applications, the Arnold conjecture."

Helmut Hofer kicked off the seminar with a chalk-and-talk abridged history: Arnold, Rabinowitz, Conley-Zehnder, Gromov, Floer. Although as we have seen, Floer theory evolved with complexity and nuance, emerging from interactions among various realms and constituents. Chance interactions created opportunity. Floer learned the right things at the right time. He took advantage of various developments in what can only be considered an audacious undertaking.

Michael Hutchings, of the University of California, Berkeley, received his Ph.D. in 1998 working with Cliff Taubes, and has followed in Floer's footsteps, after a fashion. "The biggest accomplishment of my own research is the development and application of embedded contact homology (ECH)[144] of contact three-manifolds, which can be thought of as a kind of Floer theory," he said. "The definition is more complicated than Floer's original definition but it follows in the same spirit"—it depends on the same mental framework.

Hutchings noted that the idea of Floer theory is very natural—and that it might very well have been conjured into being without Floer. "I think that if Floer hadn't existed, someone else would have thought of it," he said. "I'm just not sure how much longer it would have taken. What he did in his time apparently was very audacious, but it might not have seemed that way a few years later."

Although broadly this is true, Hofer's view is that one should reconsider the finer structure. By the 1990s, there were already the predigested versions of Floer theory, which either promoted the beautiful underlying ideas while brushing the heavy technical work under the carpet, or presented the technicalities in a more streamlined way, stressing the naturality of the construction.[145] Of course, in the wild there exist many natural yet unanticipated relationships among mathematical objects. The mathematical enterprise seeks to find and prove meaningful connections. As Eliashberg once said to Hofer:

"Good things prove themselves."

Hofer took this to mean that by finding the right definitions—the right relationships among objects—one can organize a natural construction, or proof, in such a way that the elegance of the argument is eminently convincing and its validity immediately clear. However, a "Thing" must first be found—amid the tangled mess of the vegetation in the jungle. And then said "Thing" must be identified as a "Good Thing." The availability of talent to take advantage of such an opportunity is rather sparse. A missed opportunity might result in an area developing in a different way, drifting in a divergent direction, making it less likely that the original opportunity would appear again in the same form. In the Floer story, at critical moments there were not many people who could have taken advantage of the "Good Things" that presented themselves along the way.

Gromov believed Floer theory must be wrong—in some sense out of naturality concerns. Namely, if one has symmetries the last thing you want to do is to break them, but that's precisely what was needed for Floer's theory. Gromov only believed once Bahri explained Hofer's paper on the Liusternik-Shnirelman theory for Lagrangian intersections, [**H7**]—which was based on ideas that resonated with Gromov's own thinking.[146]

Even if in hindsight everything looks natural, finding the crucial idea in the first instance is no small feat, and then having the chops to pursue it is another matter. In this case one needed to both hold and discard key pieces of knowledge. For instance, everybody who had studied critical point theory in finite or infinite dimensions believed flows to be key—a.k.a. a deformation lemma. Those who were comfortable in this area pushed the analysis to new levels (most notably Rabinowitz and his followers), but they were not interested in, or did not know, Witten's ideas. One needs to make the crucial step of forgetting about flows and replacing them with some kind of bounded solutions for certain kinds of partial differential equation. This is a big cultural step. Then one needs to develop the expertise of gluing solutions of these equations. Taubes had introduced very useful techniques in Yang-Mills theory, but one needed exposure to these technical ideas.

And indeed, after the Rabinowitz breakthrough, Conley and Zehnder started bringing variational principles into symplectic geometry. Conley was strong on ideas but less technically inclined, whereas Zehnder was an analyst at heart. They had contemplated the study of the unregularized gradient, albeit without having any idea about how to start such an approach. Gromov and later McDuff were looking into other more geometric pastures, with an abundance of low-hanging fruit. The momentum of this activity would have drawn in many researchers, but they would have speedily moved along a path that made an interaction producing something akin Floer theory less likely—at least for this group of researchers.

So, contemplating different scenarios relevant for our storyline, one wonders: Who would have been in the position to play Floer's role? The answer seems to be nobody. The development would have been considerably different. It is Hofer's guess that the appearance of a Floer-type theory would have taken substantially more time and the immediate development, after Gromov's contribution, would have gone in a more geometric direction.[147]

Good Things To Come... Over the course of the seminar at IAS, Venkatesh got a feel for Floer theory, and how complicated it is to carefully define. He also observed that Floer himself did not always carefully define it. "And that's very courageous, actually," Venkatesh said. "If I were doing that, I would get panicky. There are a tremendous number of analytic details. If Floer had sat around writing them all out—god knows how many years he would have spent. But then, without writing them out, you need to have a certain confidence that there's not some strange thing that you're missing, that you don't know."

Most of the seminar's attendees had no experience with the material. Others had waded in, at varying times and to varying degrees. Karen Uhlenbeck, a distinguished visiting professor at the Institute, who in 2019 was the first woman to win the Abel Prize, attended weekly. The subject was her mathematical milieu, but only from afar. "For me, the seminar was kind of magic," Uhlenbeck said. "Because when Floer produced this stuff, I purposely didn't learn it. I was doing other things. I recognized the mathematics, but I didn't pay attention. And the seminar picked up where I left off, starting at the beginning." Uhlenbeck lectured on one small piece of the overall picture. She knew the proof of the theorem in the textbook, but she found another proof that she thought her fellow Floer learners could better understand. "And I actually used that method in a new paper I was writing, so it benefited my research," she said.

For Shira Tanny, then a post-doc at IAS who works mainly in symplectic geometry and Hamiltonian dynamics, the seminar covered material that she partially knew—or was supposed to know, but had forgotten. Nevertheless, she said, there were interesting insights. People from different fields presented ideas or asked questions that she hadn't considered. "These questions never occurred to me," she said. "Because sometimes there are questions that you would think of from the outside, but not from the inside." Tanny also benefited from a more comprehensive historical viewpoint; some of the nuances seem more natural once she was aware of the motivations. But these days, she noted, Floer theory can hardly be thought of as one single theory or entity: "There is a core idea that is the common ground to everything that's called a Floer-type theory now, but there are lots of versions that are often quite different from one another."

Beyond its origin story, Floer theory has grown through a gigantic effort of many mathematicians. Even Floer realized that the crucial idea of counting essential orbits was much more general than the particular application for the Arnold conjecture—he applied it to a very different problem, which became known as instanton Floer homology. Other outgrowths of Floer theory include:

Lagrangian Floer homology
Fukaya categories
Rabinowitz Floer theory
Instanton homology
Symplectic homology
Seiberg-Witten Floer homology
Heegaard Floer homology
Embedded contact homology
Symplectic field theory
Floer homotopy theory

Floer's core ideas—studying infinite-dimensional spaces via the tools of Morse theory and partial differential equations—initiated over time the work of many mathematicians with a broad array of mathematical backgrounds. Entirely new fields of research and theories now called "Floer-type theories" were created and led to the resolution of longstanding problems and breakthroughs on questions which seemed to be out of reach before his seminal contributions. While Floer's initial contributions were in symplectic geometry and Yang-Mills theory, it did not take a long time before his ideas entered low-dimensional topology and the theory of dynamical systems.

In terms of Floer's broader legacy, the Floer Centre for Geometry was founded in 2011 to cultivate research in various branches of geometry at the Ruhr-Universität Bochum.

Floer homology is one of the most versatile and far-reaching tools in modern mathematics. From symplectic topology to algebraic geometry, low-dimensional topology, and mathematical physics, Floer's insights continue to inspire and impact diverse areas of mathematics. Meanwhile his ideas were also applied in group theory and quite recently in number theory. It is very rare that an idea born within a specific field becomes so important in other areas. This success should remind us of the development of "Homological Algebra." Thirty years after he had started a revolution, Henri Cartan was in awe about the success of this theory. Unfortunately, we will never know how Floer would have thought about the success story he initiated. As mathematical ideas are usually born in a specific area of mathematics, with its own questions, problems and viewpoints, the ideas reflect what one might call the "culture of the original field." If this idea is powerful enough to become of relevance in another field, we will see that it introduces new viewpoints and ultimately is morphed to be useful in the new context with a feedback to its origin. This process, which in the case of Floer theory is still ongoing, will lead to deeper understanding and broadening of the ideas since they are being looked at from many different points of view.

From this perspective one dream for the future of Floer theory is unification—a unified Floer theory, of sorts, a framework within which all the theories fit and work together. This is something that Kai Cieliebak "wasted a lot of time on" during his Ph.D. in the early-to-mid 1990s, when there were merely a couple variations. "I wanted to define a general version of Floer homology," he said. "It was completely clear that it was always the same idea. But each time, in each situation, you basically have to redo everything—you have to redo the whole analysis—and you have to write hundreds of pages to make everything work." Yet, when you look at all the different offshoots, the underlying principles are always the same. "They beg for a general, all-encompassing theory," said Cieliebak. Cieliebak spent at least a year thinking about it, but failed. "It's findable," he said. "I still believe it's findable. I haven't given up on that. If ever I have a bit of time on my hands, I might actually get back to it."

In Cieliebak's opinion, one viable path is the "polyfolds" framework developed by Hofer, Wysocki and Zehnder over the course of two decades, from 2000 to 2020, [**HWZ**]. "It is really supposed to give you the answer to what the general theory should be, but the answer has not actually been carried out. I think the answer is coming from polyfolds—polyfolds provides the correct foundations."

And circa the mid-1990s, another approach had emerged: Kuranishi structures—by Kenji Fukaya, a former senior faculty at the Simons Center for Geometry and Physics at Stony Brook, and now Professor at the Beijing Institute of Mathematical Sciences and Applications (BIMSA) and Kaoru Ono, now at Kyoto University,[**FuO**] (and later joint work with Yong-Geun Oh and Hiroshi Ohta, [$\mathbf{FO}^3_I, \mathbf{FO}^3_{II}$]).

Critiques and elaborations followed sixteen years later by McDuff, and Katrin Wehrheim, at the University of California, Berkeley,[**McWQ**]. Their work, plus that of John Pardon, of Princeton and Stony Brook, [**Pa**], provided improvements upon the original. This paved the way for recent developments that give cause for cautious optimism: most notably contributions by Mohammed Abouzaid and Andrew Blumberg, [**AbB**], Shaoyun Bai and Guangbo Xu, [**BX**], and Semon Rezchikov, [**Re**], whose works are currently being refereed. The constructions of this new work are still very involved and call for conceptual simplification. As usual, mathematics holds the opportunity for finding the right level of generality, but in this case many things have yet to be worked out.

Hutchings' current work aims at that goal—he's developing more elementary tools that can do similar things as Floer theory. "We'll see how far that goes," he said. Julian Chaidez, formerly a Ph.D. student with Hutchings, now an assistant professor at the University of Southern California, sees the value in simplification, but, alas, no easy path. "If I knew how to simplify Floer theory to something that any mathematician could get into quickly, that would be a pretty big contribution," he said. "There is a tendency for tools in mathematics to become so difficult to use that only the real experts in those tools can actually use them," Chaidez said. "You want math tools to be as simple as possible and to be as easy to use as possible. And I don't think that people have really figured out—especially with some of the more advanced stuff coming out—how to make Floer theory as user-friendly as possible."

Even as it stands, Floer theory is an enticing—perhaps irresistible, if intimidating—addition to the mathematical toolkit. "It seems absolutely terrifying, like something that's not going to end well," Venkatesh said. But it also seems like a fundamental mathematical structure—"it occurs in so many places in topology of three and four manifolds, it makes you think it's really something fundamental, like topology itself." Convening the Floer learning seminar, Venkatesh had no hidden motives in terms of his own research. Yet by the end, he had started to wonder... "I started to think, 'Oh, it will be interesting to look for analogs of those things in number theory.' It's very pie-in-the-sky," he said. "But I would like to think about it: Some of the Floer-type structures in three- and four-dimensional topology, do they have shadows in number theory? That's at least a question worth thinking about."

"It's wonderful, this Floer idea," Venkatesh said. "I'm an outsider, but that much is clear, even to an outsider. It's a wonderful thing."

THE END

APPENDIX A

Floer's Bibliography

(1) Fixed point results for symplectic maps related to the Arnold conjecture (with E. Zehnder). In: Dynamical Systems and Bifurcation, Proc. Groningen 1984, Lecture Notes in Math. 1125, Springer 1985, 47–63.

(2) Proof of the Arnold conjecture for surfaces and generalizations to certain Kähler manifolds. Duke math. J. 53 (1986), 1–32.

(3) A refinement of the Conley index and an application to the stability of hyperbolic invariant sets. Ergod. Th. and Dynam. Sys. (1987), 7, 93–103.

(4) The equivariant Conley index and bifurcations of periodic solutions of Hamiltonian systems (with E. Zehnder). Charles Conley Memorial Vol., Ergod. Th. Dynamical Systems 8* (1988), 87–98.

(5) Nonspreading wave packets for the cubic Schrödinger equation with a bounded potential (with A. Weinstein). J. Funct. Anal. 69 (1986), 397–408.

(6) Monopoles on asymptotically Euclidean 3–manifolds. Bull. Amer. Math. Soc. 16 (1987), 125–127.

(7) A topological persistance theorem for normally hyperbolic manifolds via the Conley index. Transact. Amer. Math. Soc. 321 (1990) No. 2, 647-657.

(8) Viterbo's index and the Morse index for the symplectic action, in: Periodic Solutions of Hamiltonian Systems and related topics, edited by A. Ambrosetti, I. Ekeland, P. Rabinowitz and E. Zehnder, NATO ASI Series C, Vol. 209, (1987) 147–152.

(9) Holomorphic curves and a Morse theory for exact symplectomorphisms. In: Aspect dynamiques et topologiques des group infinis de transformation de la mécanique classique, Traveaux en Cours 25, 49–61.

(10) Morse theory for fixed points of symplectic diffeomorphisms. Bull. Amer. Math. Soc. 16 (1987), 279–281.

(11) Morse theory for Lagrangian intersections. J. Diff. Geom. 28, (1988) 513–547.

(12) A relative Morse index for the symplectic action. Comm. Pure Appl. Math., Vol. XLI (1988), 393–407.

(13) The unregularized gradient flow of the symplectic action. Comm. Pure Appl. Math., Vol. XLI (1988), 775–813.

(14) Witten's complex and infinite dimensional Morse theory. Journal Differential Geom. 30 (1989), 207–221.

(15) Cuplength estimates for Lagrangian intersections. Comm. Pure Appl. Math., Vol. XLII (1989), 335–356.

(16) Selfdual conformal structures on $l\mathbf{C}P^2$. Journal Differential Geometry 33 (1991), 551-573.
(17) An instanton–invariant for 3-manifolds. Comm. Math. Phys. 118 (1988), 215–240.
(18) Symplectic fixed points and holomorphic spheres. Comm. Math. Phs. 120 (1989), 575–611.
(19) A proof of the Weinstein conjecture in $P \times \mathbf{C}^l$ (with H. Hofer and C. Viterbo). Math. Zeit. 203 (1990), 469–482.
(20) Instanton homology, surgery and knots, in Geometry of low–dimensional Manifolds, London Mathematical Society Lecture Note Series, Edited by K. Donaldson and C.B. Thomas, Vol.150, (1989), 97–114.
(21) Elliptic methods in variational problems. Proceedings of the ICM Kyoto 1990, Vol. 1, 87–97.
(22) Instanton homology. Notes by Dieter Kotschick. London Math. Soc. Lecture Note Ser., 150, Geometry of low-dimensional manifolds, 1 (Durham, 1989), 115–124, Cambridge Univ. Press, Cambridge, 1990.
(23) Instanton Homology. Springer Lecture Notes in Physics 375 (1991), 219–221.
(24) Coherent orientations for periodic orbit problems in symplectic geometry (with H. Hofer). Math. Zeit. 212 (1993), 13–38.
(25) Symplectic homology I: Open sets in $\mathbf{C}^n$ (with H. Hofer). Math. Zeit. 215 (1993), 37–88.
(26) Symplectic homology II: The general construction (with K. Cieliebak and H. Hofer). Math. Zeit. 218 (1995), 103–122.
(27) Transversality in elliptic Morse theory for the symplectic action (with H. Hofer and D. Salamon). Duke Math. J. 80 (1995), no. 1, 251–292.
(28) Applications of symplectic homology I (with H. Hofer and K. Wysocki). Math. Zeit. 217 (1994), 577–606
(29) Applications of symplectic homology II: Stability of the action spectrum (with K. Cieliebak, H. Hofer and K. Wysocki). Math. Z. 223 (1996), no. 1, 27–45.
(30) The configuration space of Yang-Mills-Higgs theory of asymptotically flat manifolds. The Floer memorial volume, 43–75, Progr. Math., 133, Birkhäuser, Basel, 1995.
(31) Monopoles on asymptotically flat manifolds, The Floer memorial volume, 3-42, Progr. Math., 133, Birkhäuser, Basel, 1995.
(32) Instanton homology and Dehn surgery, The Floer memorial volume, 77–97, Progr. Math., 133, Birkhäuser, Basel, 1995.

APPENDIX B

Qualifying Exam

Drafts of Floer's letters to the Technische Hilfswerk and his former diploma thesis advisor Ralph Stöcker indicate that he took the qualifying exam most likely in fall 1983 before he returned to Germany, where he stayed for most of 1984.

Andreas Floer
Syllabus for the qualifying examination

<u>Analysis</u>

Topological vector spaces from locally convex to Hilbert.
Frechet dericative, inverse and implicit function theorems.
Examples: Sobolev spaces and diff. operators.
Distributions, fundamental solutions of diff. op. with const. coeff.
Operator calculus, spectral theorem.
Schroedinger operators, spectral th. and quantum mechanics.
Analysis on manifolds, differential forms, pseudodiff. op.,
relation to homology

<u>Algebraic Topology and Homological Algebra</u>

Topological spaces, homology and homotopy groups.
Different definitions of homology groups: Simplicial, Chech,
singular, De Rham. Basic properties, Axioms.
Other homology theories: Stable homotopy, K-theory, K_H-theory.
Classifying spaces for cohomology theories.
Hurewicz fibrations and Fibre bundles.
Poincare spaces and manifolds.

<u>Lie Groups and Lie Algebras</u>

Definition and Relation: local Lie groups, Campbell Hausdorff,
enveloping algebra, group algebra, associated Hopf algebra.
Cohomology of Lie algebras and Lie groups.
Killing form, semisimple and simple Lie algebras.
Classification of semisimple Lie Algebras.
Classifying spaces of Lie groups, principle bundles, characteristic
classes.

<u>Special Topic</u> : Yang Mills fields.

APPENDIX C

Arbeitstagung, June 14th 1986

Programm der Mathematischen Arbeitstagung 1986 (I)

Freitag, den 13.6.1986

16.00 - 17.00 Uhr	M.F. ATIYAH: The logarithm of the Dedekind η-function

Samstag, den 14.6.1986

10.00 - 11.00 Uhr	CH. SOULÉ: Higher dimensional Arakelov theory
11.45 - 12.45 Uhr	M. KRECK: 7-dimensional Einstein manifolds with SU(3) x SU(2) x U(1) - symmetry
17.00 - 18.00 Uhr	K. FUKAYA: Collapsing and eigenvalues

Sonntag, den 15.6.1986

9.45 - 10.00 Uhr	Festlegung der nächsten Vorträge
10.00 - 11.00 Uhr	A. FLOER: Holomorphic curves and fixed points of symplectic maps
11.45 - 12.45 Uhr	P. KRONHEIMER: Gravitational instantons and Kleinian singularities
17.00 - 18.00 Uhr	G. FALTINGS: Hodge-Tate structures

FIGURE 1. Conference program from Mathematische Arbeitstagung, 1986

Titel: Holomorphic curves and fixed points of symplectic maps

Autor: Andreas Floer

Adreass: Courant Institute, 251 Mercer Street, New York, NY 10012, USA
Juli/August 86: Inst. f. Math., Ruhr Univ. Bochum, 463 Bochum.

Let P be a smooth compact manifold with a symplectic, i.e. closed and nondegenerate, 2-form ω. Then for every (time dependent) function H_t: $P \to \mathbb{R}$, one defines the (time dependent) Hamiltonian vector field X_t by

$$\omega(X_t, \cdot) = dH_t .$$

If we integrate such a family of vector fields, we obtain a family of diffeomorphisms ϕ_t of P, which preserve the symplectic form. The set of all diffeomorphisms obtained in such a way by any function H_t is called the set $\mathcal{D}_e$ of exact deformations of P.

We are interested in the fixed points of exact deformations. Clearly, the number of fixed points of $\phi \in \mathcal{D}_e$ satisfies the same estimates as the number of zeroes of a smooth vector field on P. This follows from the Lefschetz fixed point theorem, since each $\phi \in \mathcal{D}_e$ is homotopic to the identity. It has been conjectured by V.I. Arnold that due to the additional restrictions on $\mathcal{D}_e$, the number of fixed points of an exact deformation even satisfies estimates similar to those for the critical points of a smooth function on P. We have the following results in this direction:

<u>Theorem: Let (P,ω) be a compact symplectic manifold and let ϕ be an exact deformation. If all fixed points of ϕ are nondegenerate, then there number is greater than or equal to the sum of the Z_2 - Betti numbers of P.</u>

Autor: Andras Floer, **Seite:** 2

More precisely, we show that if C denotes the free Z_2-module over the set of fixed points of Φ, then there exists a homomorphism $\delta: C \to C$ satisfying $\delta\delta = 0$ and

$$H^*(P, Z_2) = \ker \delta / \operatorname{im} \delta .$$

If ϕ is a deformation induced by a time independent function H, then this follows immediately from "Morse theory", see f.e. M . In the general case, there is still a variational approatch, however one has to pass to the infinite dimensional space

$$\Omega(\phi) = \{z \in C^\infty([0,1], P) \mid z(1) = \phi(z(0))\} .$$

There is a function $\mathcal{a} : \Omega \to \mathbb{R}$ (sometimes defined only locally), which is called the symplectic action functional and which satisfies

$$d\mathcal{a}(z)\xi = \int_0^1 \omega(\xi, \dot{z}) .$$

The (nondegenerate) critical points correspond to the (nondegenerate) fixed points of $\mathcal{a}$. Since standard variational methods do not apply in this case (for reasons of lack of Palais-Smale compactness and definiteness), we use the following method: For suitable almost complex structures J on P, maps $u: \mathbb{R} \to \Omega$, $u(s)(t) = u(s,t)$ which satisfy

$$u = \frac{\partial u}{\partial s} + J\frac{\partial u}{\partial t} = 0$$

can be interpreted as trajectories of the "gradient flow" of $\mathcal{a}$. ~~Hence we are interested~~

Autor: Andreas Floer **Seite:** 3

On the other hand, they can be considered as holomorphic maps from $\mathbb{R} \times [0,1] \subset \mathbb{C}$ into (P,J). It turns out that if such a holomorphic map has finite area, then it represents a trajectory connecting two critical points. Moreover, there exists a "relative Morse index" $\mu(x,y)$ for critical points x and y so that the set of trajectories connecting x and y is a smooth manifold of dimension $\mu(x,y)$. In particular, if $\mu(x,y) = 1$, then there are is a discrete and, in fact, finite set of one dimensional trajectories. If we use these numbers as the matrix elements of the coboundary operator δ, then by establishing existence and continuation properties for holomorphic curves one can show that δ has the properties stated above. This proves the theorem.

[M] : Milnor, Lectures on the H-cobordism theorem

[W] : Witten, Supersymmetry and Morse theory

[FU] Freed/Uhlenbeck: Instantons and 4-dimensional topology

APPENDIX D

Progress Report

Progress Report Summer/Fall 1988

Andreas Floer, University of California, Berkeley

During the above period, progress was made in the following two areas:

(1) Instanton homology for 3-manifolds:

We constructed an exact sequence for instanton homology I (M) (see [F2]) of homology 3-spheres: Let K be a knot in M and let M + K be the homology 3-sphere obtained by (+1)-surgery on K. Then tere exists a long exact sequence relating I (M), I (M+K), and a new knot invariant I (K) To define the latter, one considers the manifold obtained by zero framed surgery on K, and applies the definition of instanton homology to the SO(3) - bundle with nontrivial second Stiefel- Witney class. One immediate result is that I (K) depends only on the knot complement. The construction can also be used to links, so that at least in principle, the calculation of I (M) can be reduced to the calculation of certain invariants of links in the three-sphere.

(2) Symplectic geometry:

In collaboration with H. Hofer and I. Ekeland, we gave a new definition of Ekeland-Hofer's capacity functions, which relate geometrical properties of subsets of the linear symplectic space to the existence problem for closed solutions of of Hamilton's equations on their boundaries. It uses the variational methods developed in [F1] and [F3] in the context of Lagrangian intersections and symplectic fixed point theory. The main new result here is a product formula for the capacity functions. The new definition may also be used to extend the notion of symplectic capacities to more general manifolds.

References

[EH] Ekeland, I, Hofer, H.: Convex Hamiltonian systems and their periodic trajectories, Commun. Math. Phys. 113 (1987), 419-496.

[F1] Floer, A.: Morse theory for Lagrangian intersections, J. Diff. Geom. 28,3 (1988).

[F2] —: An instanton-invariant for 3-manifolds, Commun. Math. Phys. 118 (1988), 215-240.

[F3] —: Symplectic fixed points and holomorphic spheres, to appear in Commun. Math. Phys.

APPENDIX E

The (PS) Extension

With the great success of finite-dimensional Morse theory, it was natural to study next the question of whether some of the infinite-dimensional variational principles arising in mathematics and physics could be made rigorous and useful. Morse already used one of these principles. He approximated its associated functional by suitable high-dimensional maps, studying their critical points and then drawing conclusions about the original problem. In [**Mo2**, **Mo3**] he studied the so-called geodesic problem, which allows a particularly pleasing approximation.

These ideas worked well in many cases, but mathematically it seemed more satisfying to work directly with the infinite-dimensional problem. There was reason to believe that this could be done. After all, if you can approximate an infinite-dimensional problem by a sequence of finite-dimensional problems in a way that allows us to draw conclusions, then it shouldn't be impossible to make the desired step forward. To obtain a generalization to infinite dimensions, a sufficient condition on the variational principle had to be found.

We consider a smooth function on a finite-dimensional manifold

$$f : \Lambda \to \mathbb{R}$$

and assume that Λ, equipped with a compatible metric d, is a compact space. Then the compactness implies that any given sequence on Λ has a convergent subsequence. In particular, if we find a sequence (x_k) on which the sequence of derivatives $(df(x_k))$ converges (in some norm) to 0, then there must be a critical point. Indeed, after taking a subsequence, we may assume that $x_k \to x$ for a suitable $x \in \Lambda$, and therefore

$$f(x_k) \to f(x) \quad \& \quad df(x_k) \to df(x) = 0.$$

If Λ is not compact but (Λ, d) is complete, the fact that $df(x_k) \to 0$ and $f(x_k) \to d \in \mathbb{R}$ is not enough to deduce the existence of a critical point, as the following one-dimensional example shows:

$$f : \mathbb{R} \to \mathbb{R} : s \to e^{-s}.$$

The function is positive everywhere, bounded from below by 0, but it still does not have any critical points. The derivative is $f'(s) = -e^{-s}$ and it is always strictly negative so that the function is strictly decreasing. For every given $s \in \mathbb{R}$ we have that $f(t) < f(s)$ for $t > s$. As a consequence, given any sequence of points (s_k) tending to $+\infty$ (towards the right), we find that the values of f on this sequence, i.e. $f(s_k)$, tend to 0 as $k \to \infty$ and the values of its derivative do the same:

$$f(s_k) \to 0 \quad \text{and} \quad f'(s_k) \to 0.$$

This is an example of unwanted behavior—the points s_k for large k are "almost" critical points in the sense that $f'(s_k)$ is getting closer to 0, and at the same time

$f(s_k)$ converges to a definite value, which is 0 in our case, but there is no critical point at this level. The last example is ruled out if we require the following for a function $f : \Lambda \to \mathbb{R}$ defined on a smooth finite-dimensional manifold equipped with a complete Riemannian metric:

(*) Any sequence $(x_k) \subset \Lambda$ satisfying $\|df(x_k)\| \to 0$ and $f(x_k) \to d \in \mathbb{R}$ has a convergent subsequence.

This immediately implies that there exists $x \in \Lambda$ with $f(x) = d$ and $df(x) = 0$. We note that $s \to e^{-s}$ does not satisfy this condition.

The condition (*) is deceptively simple. If a functional f satisfies it, we only need methods for finding sequences (x_k) for which $(f(x_k))$ converges to some real number and $(df(x_k))$ converges to 0 in order to find critical points. It is not immediately obvious that the condition guarantees a relationship between the structure of critical points and some count of holes. However, this relationship might look different from that in the case when Λ is compact, since in the non-compact case the behavior of f at infinity matters as well.

The Palais-Smale condition (PS), which was inspired by examples of the type we have discussed, just formulates our previous condition in the context of finite and infinite-dimensional spaces:

(PS). A smooth function $f : \Lambda \to \mathbb{R}$ on a smooth space Λ equipped with a complete Riemannian metric and associated norm $\|\cdot\|$ and metric ρ, satisfies the Palais-Smale condition provided the following holds. If there exists a sequence (x_k) in Λ with the property that

$$f(x_k) \to d \quad \text{and} \quad \|df(x_k)\| \to 0,$$

then there exists a point $x \in \Lambda$ and a sequence of natural numbers k_j tending to infinity, written as $k_j \to \infty$, such that the so-called subsequence (x_{k_j}) of (x_k) tends to x, i.e.

$$\rho(x_{k_j}, x) \to 0 \quad \text{as} \quad j \to \infty.$$

Above $\|df(x_k)\|$ is a certain measurement of the size of $df(x_k)$, which for our purposes is not that relevant—for example $\|\cdot\|$ is a function which defines for every $x \in \Lambda$ on the tangent space $T_x\Lambda$ a norm.

If **(PS)** holds, one can show that $df(x) = 0$ and $f(x) = d$, which means x has to be a critical point ($df(x) = 0$) at level d ($f(x) = d$). The Palais-Smale condition rules out the bad behavior illustrated by our simple example and allows one to deduce from the existence of a sequence of almost critical points, say $f(x_k) \to d$ and $\|df(x_k)\| \to 0$, that there exists a critical point x of f with $f(x) = d$.

The Palais-Smale condition simply prohibits the unwanted phenomenon that "topology" is leaking out of the space at "critical points at infinity". In applications to partial differential equations it might be quite tricky to show that the condition actually holds—it is not always true or only true for certain $d \in \mathbb{R}$.

In situations where the Palais-Smale condition holds, one was subsequently able to carry out the Morse-theoretic ideas, but only where the critical points had finite Morse indices—here one could quite often establish a relationship between global behavior of a space, behavior of f near infinity, and the properties of the critical points of the Morse-function. Without going into details now, the assumption of finite Morse indices ruled out phenomena related to our discussion in Appendix F

Starting in the 1970s, these ideas became quite popular in nonlinear analysis, where the goal was the solution of certain types of partial differential equations by variational methods. See for example the pioneering work of Antonio Ambrosetti and Rabinowitz [**AR**] and Michael Struwe's important book [**St**].

APPENDIX F

There's a Lot of Room in Infinite Dimensions

A basic result which holds for finite-dimensional Euclidean spaces is Brouwer's fixed point theorem. It says the following where we denote by D^n the closed unit ball in $\mathbb{R}^n$:

THEOREM (Brouwer). If $f : D^n \to D^n$ is a continuous map then there exists a point $z \in D^n$ satisfying $f(z) = z$.

Infinite-dimensional Hilbert spaces—generalizing the finite-dimensional Euclidean spaces—like $\ell_2 := \ell_2(\mathbb{N})$ have interesting properties which are impossible in finite dimensions, and sometimes defy intuition. The elements of the space $\ell_2(\mathbb{N})$ are all the sequences $(z_1, z_2, \cdots)$ of real numbers satisfying $\sum_{j=1}^{\infty} |z_j|^2 < \infty$. One can define a norm and a distance by

$$\|z\| = \sqrt{\sum_{j=1}^{\infty} |z_j|^2} \quad \& \quad d(z, z') = \|z - z'\|.$$

We denote by O the zero sequence and by $\mathbf{D}$ the collection of all points z in ℓ_2 with distance at most 1 to O and by $\mathbf{S}$ its boundary, which consists of all z with distance to O precisely 1. Hence $\mathbf{D}$ is the closed unit ball and $\mathbf{S}$ the closed unit sphere in ℓ_2. View $\mathbf{D}$ as an infinite-dimensional soccer ball with its surface being $\mathbf{S}$.

We have already colloquially mentioned before that there is a lot of room in infinite dimensions. The prototype "exhibit" here is the shift map

$$T : \ell_2 \to \ell_2 : T(z) = (0, z_1, z_2, \ldots).$$

This map fixes O and preserves distances between points since $\|T(x) - T(y)\| = \|x - y\|$, but its image which consists of all z with $z_1 = 0$ lost one dimension. Applying it again we lose another dimension.

Compare this with the finite-dimensional situation. In finite dimensions an isometry of the Euclidean (affine) space, i.e. a map of $T : \mathbb{R}^N \to \mathbb{R}^N$ which has the property

$$d_{\text{Euclid}}(T(z), T(z')) = d_{\text{Euclid}}(z, z') \text{ for all } z, z',$$

has the form $T(z) = \bar{z} + A(z)$, where $\bar{z} \in \mathbb{R}^N$ is fixed and A is an orthogonal matrix

$$\langle A(x), A(y) \rangle = \langle x, y \rangle.$$

If in addition $T(0) = 0$, we see that $T(x) = A(x)$. In particular T is a linear isomorphism (For details see [**Ber**], Proposition 9.1.3.), a fact which does not hold in infinite dimensions.

We shall see next that there is no version of Brouwer's fixed point theorem for $\mathbf{D}$ by constructing a continuous map $f : \mathbf{D} \to \mathbf{D}$ without fixed points. The procedure involves several steps.

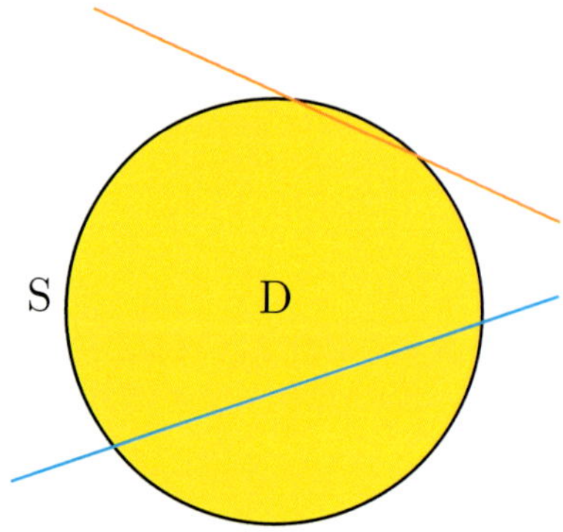

FIGURE 1. A line through two different points in **D** (yellow) intersects **S** (black boundary) precisely twice.

We first produce a continuous auxiliary map $g : \mathbf{D} \to \mathbf{S}$ which associates to a point z in **D** the point $g(z)$ in **S** defined by

$$g(z) = \left(\sqrt{1 - \|z\|^2}, z_1, z_2, z_3, \cdots\right).$$

We have shifted the sequence z one to the right and inserted a specific expression in the now empty first slot so that

$$\|g(z)\| = 1 \quad \text{for all } z \in \mathbf{D},$$

which means that $g(z)$ belongs to **S** as claimed. Also continuity of g is easily established.

The map g does not have a fixed point. To see this, assume that $g(z) = z$ which implies z is in **S**, i.e. $\|z\| = 1$ and hence $g(z)$ has the form

$$g(z) = (0, z_1, z_2, z_3, \cdots).$$

Armed with this knowledge, we go back to the equality $g(z) = z$ which tells us that

$$(0, z_1, z_2, z_3, \cdots) = (z_1, z_2, z_3, \cdots)$$

and we can read off by comparing coordinates

$$z_1 = 0, \ z_2 = z_1, \ z_3 = z_2, \ \cdots$$

implying that all z_i are 0. Hence $z = O$, which gives a contradiction since O is not in **S** and z is in **S**. This shows that g does not have fixed points and as a consequence, for $z \in \mathbf{D}$, the points z and $g(z)$ are always different. We also know that $g(z)$ lies in **S**. This allows for a further construction.

Since z and $g(z)$ are two different points they span a line, see Figure 1, and this line intersects **S** twice, because both points belong to **D** but are different, see Figure 2.

Of course, one of the intersection points is $g(z)$ given that it belongs to **S** and the line. We denote the other intersection point by $f(z)$.

The map $f : \mathbf{D} \to \mathbf{S}$ has the property that $f(z) = z$ for $z \in \mathbf{S}$. We can also define the continuous map

$$\Theta : [0, 1] \times \mathbf{D} \to \mathbf{D} : \Theta(t, z) = (1 - t) \cdot z + t \cdot f(z),$$

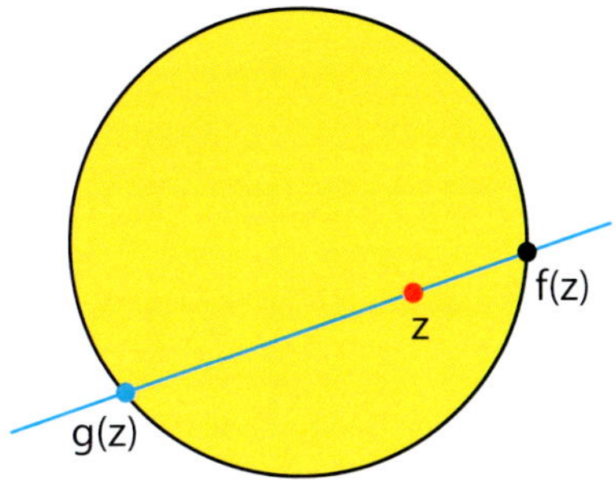

FIGURE 2. The definition of f as an intersection point with $\mathbf{S}$.

which we can view as a continuous family of maps

$$\Theta_t : \mathbf{D} \to \mathbf{D} : \Theta_t(z) := \Theta(t, z).$$

If $t = 0$ we have that $\Theta_0(z) = z$ for all $z \in \mathbf{D}$, whereas $\Theta_1(z) = f(z)$ for all $z \in \mathbf{D}$. Hence, in a continuous way, the identity map Θ_0 is deformed into f while t varies from 0 to 1. For $z \in \mathbf{S}$ we have that $f(z) = z$ implying that $\Theta_t(z) = z$ for $t \in [0, 1]$. In other words, the points in $\mathbf{D}$ are transported continuously to $\mathbf{S}$ in such a way that the points in $\mathbf{S}$ never move, but since Θ is continuous, no tearing ever happens. In finite dimensions, this is impossible by Brouwer's theorem.

We can also define the map

$$\Gamma : [0, 1] \times \mathbf{S} \to \mathbf{S} : \Gamma(t, z) = f((1 - t)z),$$

which is a continuous family of maps with $\Gamma(0, z) = f(z) = z$ for $z \in \mathbf{S}$ and $\Gamma(1, z) = f(0)$. Mathematicians call a map like Γ a homotopy of the identity map to a constant map; $\mathbf{S}$ is contractible in itself to a point, something which is not true for finite-dimensional spheres. This is the deeper reason that algebraic topology in finite dimensions, but not necessarily in infinite dimensions, is interesting and intriguing.

The fact that infinite dimensional spaces can have finite-dimensional holes allows one to extend Morse theory, as long as the critical points have finite Morse indices and certain technical compactness properties are met. The fact that these infinite-dimensional spaces do not have infinite-dimensional holes looks like the kiss of death for any attempt to generalize Morse theory to the case of infinite Morse indices.

APPENDIX G

Analysis of $\mathcal{A}_{\mathbb{R}^{2n}}$

When investigating the nature of $\mathcal{A}_{\mathbb{R}^{2n}}$ it becomes clear that in the background there is an integration of a differential one-form λ over the loop. One can take

$$\lambda = \sum_{i=1}^{n} q_i \cdot dp_i,$$

where, as before, we take the coordinates $z = ((q_1, p_1), \ldots, (q_n, p_n))$ and it can be verified that we have the identity

$$\mathcal{A}_{\mathbb{R}^{2n}}(z) = \int z^* \lambda. \tag{39}$$

Differential forms play an outsized role in geometry, topology and physics and this might be a good moment to develop familiarity with this important topic (see for example: [**AbM**, **BT**, **MaT**, **Wa**]). Reading up on things one doesn't know is an important part of research and here is a salient point: Part of increasing one's mathematical surface area is to be aware that certain types of problems are being studied in specific areas of mathematics. Even if one is far away from knowing details, this knowledge gives a starting point. In a research setting one usually studies a specific problem and it will provide a context so that one does not learn something new within a 'vacuum.' Also, the readers who are not familiar with the theory of differential forms might have seen special instances—the fundamental theorem of calculus, as well as the divergence theorem and the theorems of Green and Stokes, known from multivariable calculus, are special cases of a generalized Stokes theorem in the theory of differential forms.

The right-hand side in (39) is by definition given by a specific integral and we carry out a partial integration to derive the definition of the symplectic form

$$\begin{aligned} \mathcal{A}_{\mathbb{R}^{2n}}(z) &= \sum_{i=1}^{n} \int_{S^1} q_i(t)\dot{p}_i(t)dt \quad \left(\text{Definition of } \int z^*\lambda\right) \\ &= \frac{1}{2} \cdot \sum_{i=1}^{n} \int_{S^1} q_i(t)\dot{p}_i(t)dt + \frac{1}{2} \cdot \sum_{i=1}^{n} \int_{S^1} q_i(t)\dot{p}_i(t)dt \\ &= \frac{1}{2} \cdot \sum_{i=1}^{n} \int_{S^1} q_i(t)\dot{p}_i(t)dt - \frac{1}{2} \cdot \sum_{i=1}^{n} \int_{S^1} p_i(t)\dot{q}_i(t)dt \\ &= \int_{S^1} \frac{1}{2} \left(\sum_{i=1}^{n} [q_i(t)\dot{p}_i(t) - p_i(t)\dot{q}_i(t)] \right) dt. \end{aligned} \tag{40}$$

The last integral can be rewritten so that we obtain

$$\mathcal{A}_{\mathbb{R}^{2n}}(z) = \int_{S^1} \frac{1}{2}\omega(z(t), \dot{z}(t))dt, \tag{41}$$

where ω denotes the non-degenerate skew-symmetric bilinear form defined by

$$\begin{aligned} \omega &: \mathbb{R}^{2n} \times \mathbb{R}^{2n} \to \mathbb{R} \\ \omega(e, f) &= \sum_{i=1}^{n} (h_i \cdot k_i' - k_i \cdot h_i') \end{aligned} \tag{42}$$

with $e = (h_1, k_1, \ldots, h_n, k_n)$ and $f = (h_1', k_1', \ldots, h_n', k_n')$ in $\mathbb{R}^{2n}$. Using the calculus of differential forms one can show, first of all, that ω can be written as

$$\omega = \sum_{i=1}^{n} dq_i \wedge dp_i \tag{43}$$

and, moreover, that ω is the exterior derivative of the one-form $\lambda = \sum_{i=1}^{n} q_i dp_i$

$$\omega = d\lambda.$$

The two-form ω is non-degenerate and since $\omega = d\lambda$ it follows that $d\omega = d^2\lambda = 0$ since by an important property of the exterior derivative it holds that $d \circ d = 0$. One says that ω is a closed non-degenerate two-form on $\mathbb{R}^{2n}$. Moreover $(\mathbb{R}^{2n}, \omega)$ is by definition a symplectic vector space. More generally:

DEFINITION. *A symplectic vector space (V, ω) consists of an even-dimensional real vector space V and a non-degenerate skew-symmetric bilinear form ω defined on V.*

One can formally differentiate $\mathcal{A}_{\mathbb{R}^{2n}}$ as follows. We pick loops z and h and would like to take the derivative of $\mathcal{A}_{\mathbb{R}^{2n}}$ at z into the direction h, which we shall denote by $d\mathcal{A}_{\mathbb{R}^{2n}}(z)(h)$. For this let $\tau \in (-1, 1)$ and consider the path γ in $\text{Loop}(\mathbb{R}^{2n})$ by

$$\gamma(\tau) = z + \tau h.$$

For fixed τ our loop is $z + \tau h$, which is given by $S^1 \ni t \to z(t) + \tau \cdot h(t)$. The path γ is a family of loops parametrized by τ and therefore we are able to consider the function

$$(-1, 1) \to \mathbb{R} : \tau \to \mathcal{A}_{\mathbb{R}^{2n}}(\gamma(\tau)).$$

Now we are in the world of one-variable calculus and compute the derivative at $\tau = 0$. We obtain

$$d\mathcal{A}_{\mathbb{R}^{2n}}(z)(h) = \frac{d(\mathcal{A}_{\mathbb{R}^{2n}}(\gamma(\tau)))}{d\tau}_{|\tau=0} = \int_{S^1} \omega\left(h(t), \frac{dz}{dt}(t)\right) \cdot dt.$$

Note here that $\mathcal{A}_{\mathbb{R}^{2n}}(\gamma(\tau))$ is an expression of the form

$$A + B\tau + C\tau^2.$$

We leave it to the reader to find the integral expressions for A, B and C. The derivative at $\tau = 0$ for this type of expression is then B.

Doing this for fixed z in all directions h we see that the derivative $d\mathcal{A}_{\mathbb{R}^{2n}}(z)$ vanishes on h precisely when $\frac{dz}{dt} = 0$, which means z is a constant loop. At least formally z is a critical point of $\mathcal{A}_{\mathbb{R}^{2n}}$ if and only if z is a constant loop. Of course, that does not seem to be such an exciting fact.

Assume next that a smooth map $H : \mathbb{R}^{2n} \to \mathbb{R}$ is given so that we can build the new functional

$$\operatorname{Loop}(\mathbb{R}^{2n}) \to \mathbb{R} : z \to \mathcal{A}_{\mathbb{R}^{2n}}(z) - \int_{S^1} H(z(t)) \cdot dt.$$

Thinking of $\mathbb{R}^{2n}$ as the space of states for some physical system, we may think of H as an energy function, which assigns to a state an energy. Our functional assigns to a loop of states the difference of the action of the loop and the mean value of the energy over the loop.

Formally, what are the critical points of our new functional? A computation shows that the derivative of the new functional vanishes if and only if

$$\int_{S^1} \left(\omega \left(h(t), \frac{dz}{dt}(t) \right) - dH(z(t))(h(t)) \right) \cdot dt = 0 \quad \text{for all } h. \tag{44}$$

We can write $dH(z)(h)$ as

$$\sum_{j=1}^{n} \frac{\partial H}{\partial q_j}(z) \cdot a_j + \frac{\partial H}{\partial p_j}(z) \cdot b_j,$$

where h is given by

$$h = \begin{bmatrix} a_1 \\ b_1 \\ \cdot \\ \cdot \\ a_n \\ b_n \end{bmatrix}$$

and abbreviate the so-called gradient of H by

$$\nabla H(z) = \begin{bmatrix} \frac{\partial H}{\partial q_1}(z) \\ \frac{\partial H}{\partial p_1}(z) \\ \cdot \\ \cdot \\ \frac{\partial H}{\partial q_n}(z) \\ \frac{\partial H}{\partial p_n}(z) \end{bmatrix}.$$

Now we can rewrite (44) as

$$\int_{S^1} \left(\omega \left(h(t), \frac{dz}{dt}(t) \right) - \langle h(t), \nabla H(z(t)) \rangle \right) \cdot dt = 0 \quad \text{for all } h. \tag{45}$$

Since the expression in (44) vanishes for all h, it follows after a short computation writing $z_i = (q_i, p_i)$, $z = (z_1, \dots, z_n)$:

$$\begin{aligned} \dot{q}_1(t) &= \frac{\partial H}{\partial p_1}(z) \ \& \ \dot{p}_1(t) = -\frac{\partial H}{\partial q_1}(z) \\ &\cdots\cdots\cdots \\ \dot{q}_n(t) &= \frac{\partial H}{\partial p_n}(z) \ \& \ \dot{p}_n(t) = -\frac{\partial H}{\partial q_n}(z). \end{aligned} \tag{46}$$

These are the famous Hamiltonian equations in the special case of $\mathbb{R}^{2n}$.

APPENDIX H

Arnold's Motivation

The motivation for Arnold's conjecture: In a nutshell, if the Hamiltonian is autonomous—it does not depend on time—the Arnold conjecture follows at once from Morse theory, since the critical points of H are periodic orbits. If the time-dependent Hamiltonian is small enough, it is not too difficult to show that the number of periodic orbits associated to the Hamiltonian system can be estimated again in the non-degenerate case by Morse theory for an auxiliary problem. Arnold claimed that this estimate is still true without the smallness condition. This is quite daring, since all ideas at the time (and there weren't many) collapse instantaneously.

PROPOSITION. *Let (M, ω) be a compact symplectic manifold. Then for every smooth time-dependent Hamiltonian $H : [0,1] \times M \to \mathbb{R}$ which is C^2-small, the time-one map Ψ of the associated Hamiltonian system has the property that there exists a smooth map $f : M \to \mathbb{R}$ so that the fixed points of Ψ are precisely the critical points of f. If Ψ has only non-degenerate fixed points*[148]*, then f is a Morse function.*

In other words, under a smallness assumption the Arnold conjecture is related to critical point theory, and under a non-degeneracy assumption, to Morse theory. The Arnold conjecture simply states that the smallness assumption is not needed.

Next we shall get into some explanatory details, which require knowledge of differential geometry. Readers without knowledge of differential geometry might read up on the necessary material; for example it can be found in [**AbM**].

Given a smooth manifold M we can take its cotangent bundle T^*M, where we write p for the points in T^*M and we denote by $\tau : T^*M \to M$ the obvious projection. Given p we abbreviate $q = \tau(p)$. The manifold T^*M admits a canonical one-form Θ—called the Liouville form—defined as follows, where $h \in T_p(T^*M)$:

$$\Theta(p)(h) = p(T\tau(p)(h)).$$

Then one can show that $-d\Theta$ defines a symplectic form on T^*M. The minus sign is a convention and the construction $M \to (T^*M, \Theta)$ is canonical and requires no choices of auxiliary structures.

Then $(T^*M, -d\Theta)$ is a symplectic manifold associated to a given smooth manifold M. These symplectic manifolds are very useful in classical mechanics.

It is useful to know that given a section λ of $T^*M \to M$ it holds that

$$\lambda^*\Theta = \lambda.$$

In particular we have that

$$d\lambda = d(\lambda^*\Theta) = \lambda^*(d\Theta).$$

More generally let us recall:

DEFINITION. *A symplectic manifold is a pair* (W, ω), *where* $W = W^{2n}$ *is a smooth manifold and* ω *is a closed non-degenerate 2-form, i.e.* $d\omega = 0$ *and* ω^n *is a volume form.*

An important classical result is Darboux's theorem, which plainly states that locally symplectic manifolds of equal dimensions look the same. This is a quite different situation from that of Riemannian manifolds.[149]

THEOREM (Darboux). *Given two symplectic manifolds* (M, ω) *and* (N, τ) *of the same dimension and points* $m \in M$ *and* $n \in N$, *there exists a diffeomorphism between suitable open neighborhoods*

$$(U(m), m) \xrightarrow{\Phi} (V(n), n)$$

preserving the symplectic forms, i.e. $\Phi^*\tau = \omega|U(m)$ *and mapping* m *to* n.

As a consequence we find, for a given (M, ω) around every point $m \in M$, a chart with its image in the standard symplectic $\mathbb{R}^{2n}$ pulling back its form to ω near m. Therefore, symplectic manifolds of the same dimension look locally alike. However, we have infinitely many invariants to say how big neighborhoods are, so there is some nontrivial local information.[150]

Given a symplectic manifold (W, ω) of dimension $2n$, there is an important class of special submanifolds.

DEFINITION. *A Lagrangian submanifold of* (W^{2n}, ω) *is a submanifold* L^n *such that, with* $\iota : L \to W$ *being the inclusion, it holds that* $\iota^*\omega = 0$.

Lagrangian manifolds are half-dimensional submanifolds such that the actions of all small loops lying in them vanish.

The following result compares the neighborhood of a closed Lagrangian manifold with the neighborhood of the zero-section in its cotangent bundle. The basic result in this direction is known as Weinstein's Lagrangian neighborhood theorem:

THEOREM (Weinstein). *Given a (closed) symplectic manifold* (W, ω) *and a closed Lagrangian submanifold* L, *there exist open neighborhoods* $U(L) \subset W$ *and* $V(L) \subset T^*L$ *and a diffeomorphism* $\Phi : U(L) \to V(L)$, *which is the identity on* L *and such that* $\Phi^*(-d\Theta) = \omega|U(L)$.

DEFINITION. *The group of symplectic diffeomorphisms* $\mathrm{Symp}(W, \omega)$ *is the subgroup of* $\mathrm{Diff}(W)$ *consisting of the elements* f *with* $f^*\omega = \omega$.

Assuming for simplicity that (W, ω) is closed, we equip $W \times W$ with

$$\omega \ominus \omega := \pi_1^*\omega - \pi_2^*\omega,$$

where $\pi_i : W \times W \to W$ are the two projections.

Note that the diagonal

$$\Delta_W = \{(w, w) \mid w \in W\},$$

and the graphs of symplectic diffeomorphisms, defined by

$$\mathrm{graph}(\Phi) = \{(w, \Phi(w)) \mid w \in W\},$$

are Lagrangian submanifolds of $W \times W$.

Assume that we are given $L := \mathrm{graph}(\Phi)$ and $K := \mathrm{graph}(\Psi)$, where Ψ is C^1-close (enough) to Φ. Using the Weinstein neighborhood theorem, after a change of coordinates, we can view

$$K \subset T^*L.$$

Moreover, with Ψ close enough to Φ, K is a graph over L—that is, the image of a section of $T^*L \to L$. In other words, K is the graph of a one-form λ, and we have the embedding

$$\lambda : L \to T^*M,$$

which has image K. Since K is Lagrangian we infer that

$$0 = \lambda^*(-d\Theta) = -d(\lambda^*(\Theta)) = -d\lambda.$$

This means that the graph of a one form λ in T^*M is Lagrangian if and only if λ is closed. Of course, it is possible that λ is exact, in which case we would call the graph an exact Lagrangian graph.

A consequence of this is that the structure of $\mathrm{Symp}_0(W, \omega)$—the connected component of the identity map for a closed W—is quite nice. The local structure near an element is described by the Lagrangian picture, using closed 1-forms. In particular, $\mathrm{Symp}_0(W, \omega)$ is path-connected, since it is connected and locally path connected.

Let (W, ω) be a symplectic manifold, and for simplicity assume that W is closed. Consider a smooth map $H : [0, 1] \times W \to \mathbb{R}$, which we call in our context a (smooth time-dependent) Hamiltonian. We can associate to H a time-dependent vector field X_{H_t} by

$$-dH_t = i_{X_t}\omega.$$

By solving

$$\frac{dx}{dt}(t) = X_t(x(t)),$$

we obtain a smooth arc $t \to \Phi_t^H$ of symplectic diffeomorphisms starting for $t = 0$ at Id. Given the local structure of $\mathrm{Symp}(W, \omega)$, we can say the following: If one can reach Φ as the time-1 map of a time-dependent Hamiltonian system $K : [0, 1] \times W \to \mathbb{R}$, then we can also reach it using a 1-periodic Hamiltonian $H : S^1 \times W \to \mathbb{R}$. This implies that certain fixed points can be identified with 1-periodic orbits.

DEFINITION. *We obtain as time-1-maps the subgroup* $\mathrm{Ham}(W, \omega)$ *of*

$$\mathrm{Symp}_0(W, \omega) \subset \mathrm{Symp}(W, \omega)$$

of so-called Hamiltonian diffeomorphisms.

In the local picture, using Lagrangians, a path in $\mathrm{Ham}(W, \omega)$ connecting Id with Ψ would be locally described by a path of exact one-forms.

If $H : S^1 \times W \to \mathbb{R}$ is C^2-small, then Φ_t stays C^1-close to Id. This implies that $\mathrm{graph}(\Phi_t)$ stays C^1-close to $\Delta := \Delta_W \subset (W \times W, \omega \ominus \omega)$, and we can view, by Weinstein's theorem, $\mathrm{graph}(\Phi_t)$ as lying in a neighborhood of the zero section of $T^*\Delta \simeq T^*W$. We know that $\mathrm{graph}(\Phi_1)$ is a graph of an exact one-form.

Hence the zeros of the one-form are precisely the critical points of a function $f : W \to \mathbb{R}$. If H is C^2-small, we see that the fixed points of Φ_1 are in one-to-one correspondence to the critical points of a smooth function f on W. Namely, $\mathrm{graph}(\Phi_1) \cap \Delta_M$ precisely corresponds to the zeros of df, which are also the fixed points of Φ_1.

LEMMA. *If the fixed points of* Φ_1 *are non-degenerate in the sense that* $\mathrm{Id}_w - T\Phi_1(w) : T_wW \to T_wW$ *is invertible, then* f *is a Morse-function.*

We therefore see that if $\Phi \in \mathrm{Ham}(W, \omega)$ is generated by a C^2-small Hamiltonian, then the critical points are in one-to-one correspondence with the critical points of a smooth function on M. If Φ is non-degenerate, it is a Morse function.

Arnold published various formulations of this conjecture. Of particular interest is a formulation found in his classical book [**Ar1**][151]:

THEOREM. Every symplectic diffeomorphism Ψ of a compact symplectic manifold, homologous to the identity, has at least as many fixed points as a smooth function on this manifold has critical points (at least if this diffeomorphism is not too far from the identity).

This so-called theorem requires some explanation. As phrased, Arnold talks about Hamiltonian diffeomorphisms. In the case that the Hamiltonian diffeomorphism Ψ is possibly degenerate, the estimate is $\sharp\,\mathrm{Fix}(\Psi) \geq \sharp(M)$; and if it is non-degenerate, $\sharp\,\mathrm{Fix}(\Psi) \geq \sharp_{\mathrm{Morse}}(M)$.

At the time of Arnold's writing, without that last parenthetical bit, the statement was not known to be true. With his "not too far" smallness assumption, the statement was more or less trivial. This was classic Arnold: messing with the reader's mind.

APPENDIX I

Details for the Magic

We shall show how to mathematically encode the structure of critical points and the rigid flow lines between them.

I.1 Encoding Critical Points. We assume we are given a Morse function $f : \Lambda \to \mathbb{R}$ on a smooth closed manifold Λ of dimension L. The central features of our mathematical construction are the critical points with their Morse indices and the rigid gradient lines. We decompose the set of critical points according to their Morse indices

$$\mathrm{Crit}(f) = \mathrm{Crit}_0(f) \sqcup \ldots \sqcup \mathrm{Crit}_L(f). \tag{47}$$

We would like to construct a diagram analogous to Diagram (15) using the critical points.

The picture to have in mind is that the information on layer k is constructed as follows. Take the critical points of Morse index k and think of them as lights. A light can be lit or not, and we can describe any possible scenario for $\mathrm{Crit}_k(f)$ by a map $\sigma : \mathrm{Crit}_k(f) \to \mathbb{Z}_2$. If σ assigns to a critical point the number 1, it is lit; otherwise it is not lit. Hence, for every $k \in \{0, \ldots, L\}$ the state of $\mathrm{Crit}_k(f)$ is given by a map

$$\sigma : \mathrm{Crit}_k(f) \to \mathbb{Z}_2. \tag{48}$$

The set of all possible states on $\mathrm{Crit}_k(f)$ is just the collection of all maps as in (48), denoted by $\mathrm{Map}_k(f)$, i.e.,

$$\mathrm{Map}_k(f) = \{\sigma : \mathrm{Crit}_k(f) \to \mathbb{Z}_2\}. \tag{49}$$

The sets $\mathrm{Map}_k(f)$ allow several operations. Namely, given two maps σ and τ in $\mathrm{Map}_k(f)$, we can define an addition of them as follows. The map $\sigma+\tau : \mathrm{Crit}_k(f) \to \mathbb{Z}_2$ is defined by

$$(\sigma + \tau)(x) := \sigma(x) + \tau(x) \text{ for } x \in \mathrm{Crit}_k(f), \tag{50}$$

where we note that, on the right hand side, for every x the values $\sigma(x)$ and $\tau(x)$ belong to $\mathbb{Z}_2$ and are added up as previously defined in (19).

The Abelian group $(\mathbb{Z}_2, +)$ can be enhanced to a 'field' by defining a multiplication by

$$0 \cdot 0 = 0 \cdot 1 = 1 \cdot 0 \;\;\&\;\; 1 \cdot 1 = 1.$$

The multiplication is compatible with the addition, and we leave it to the reader to write down the properties—$(\mathbb{Z}_2, +, \cdot)$ is a (mathematical) field.

We can also multiply σ by a number $\varepsilon \in \mathbb{Z}_2$

$$(\varepsilon \cdot \sigma)(x) = \varepsilon \cdot \sigma(x) \text{ for } x \in \mathrm{Crit}(f). \tag{51}$$

The zero map $O \in \mathrm{Map}_k(f)$ assigns to every critical point in $\mathrm{Crit}(f)$ the number $0 \in \mathbb{Z}_2$, i.e., no light whatsoever.

In view of the properties (50) and (51), mathematicians say that $\mathrm{Map}_k(f)$ has the structure of a finite-dimensional vector space over the field $\mathbb{Z}_2$. In undergraduate linear algebra, one usually considers vector spaces over the real numbers. However, as we have just seen, the concept of a field is much more general, and one can consider vector spaces over any field. The dimension of our vector space over $\mathbb{Z}_2$ is precisely $\sharp\,\mathrm{Crit}_k(f)$, the number of critical points. A natural vector space basis for $\mathrm{Map}_k(f)$ can be given as follows: Define for every critical point x in $\mathrm{Crit}_k(f)$ the map $\delta_x \in \mathrm{Map}_k(f)$ by

$$\delta_x(y) = \begin{bmatrix} 0 & \text{if } y \neq x \\ 1 & \text{if } y = x. \end{bmatrix}$$

The state δ_x is easily explained—the light is off everywhere except at x. Then, using the vector space operations defined in (50) and (51), every σ in $\mathrm{Map}_k(f)$ can be written uniquely in the form

$$\sigma = \sum_{x \in \mathrm{Crit}_k(f)} \varepsilon_x \cdot \delta_x, \tag{52}$$

where the $\varepsilon_x \in \mathbb{Z}_2$ are uniquely determined by σ.

At this point, we have encoded the data of interest associated with f—its critical points and their Morse indices, corresponding to Diagram (15), but without the arrows.

Having associated to f the $\mathbb{Z}_2$-vector spaces $\mathrm{Map}_k(f)$, we note that we can associate to f the polynomial P_f defined by

$$P_f(t) := \sum_{k \in \mathbb{Z}} \dim_{\mathbb{Z}_2} M_k(f) \cdot t^k \quad \text{(This is a finite sum from 0 to } L.) \tag{53}$$

which is, of course, the Poincaré polynomial associated with f.

I.2 Encoding Rigid Flow Lines. For every $k \in \{1, \ldots, L\}$ we define the linear map

$$\partial_k : \mathrm{Map}_k(f) \to \mathrm{Map}_{k-1}(f) \tag{54}$$

by suitably counting rigid flow lines. Linearity means that it has the property $\partial_k(\sigma + \tau) = \partial_k(\sigma) + \partial_k(\tau)$ and $\partial_k(\varepsilon \cdot \tau) = \varepsilon \cdot \partial_k(\tau)$ for $\sigma, \tau \in \mathrm{Map}_k(f)$ and $\varepsilon \in \mathbb{Z}_2$. Note that ∂_k depends on the choice of the generic Riemannian metric g, but we shall suppress this in our notation, i.e. f stands for (f, g).

Given $\sigma \in \mathrm{Map}_k(f)$, we define the new map $\partial_k\sigma$ in $\mathrm{Map}_{k-1}(f)$ as follows: For a critical point $x \in \mathrm{Crit}_{k-1}(f)$, define for y with $m(y) = m(x) + 1$ the number $\chi(x, y) \in \mathbb{Z}_2$, which is 0 if the number of rigid trajectories from y to x is even, and 1 if this number is odd. The map $\partial_k\sigma \in \mathrm{Map}_{k-1}(f)$ is then defined by

$$(\partial\sigma)(x) = \sum_{\{y \in \mathrm{Crit}(f) \,\mid\, m(y) = m(x)+1\}} \chi(x, y) \cdot \sigma(y), \quad x \in \mathrm{Crit}_{k-1}(f). \tag{55}$$

This formula does precisely what we described when discussing Witten's idea. Given a state σ, the new state $\partial_k\sigma$ describes what the communication channels (after superposition) indicate. Namely given a critical point x the value $(\partial_k\sigma)(x)$ is obtained from the $\sigma(y)$ with $m(y) = m(x) + 1$ as follows. First we take the $\mathbb{Z}_2$-count of rigid flow lines from y to x and denote it by $\chi(x, y)$. This number is 0 if there is an even number from y to x or 1 otherwise. If the lamp at y is switched on, $\sigma(y) = 1$, each flow line from y to x reports a 1 and taking the superposition the contribution coming from $\sigma(y)$ is $\chi(x,y)\sigma(y)$. The same holds if $\sigma(y)=0$. Then we take the superposition at x by varying over y resulting in the formula (55). In other words, if σ is a given state, then $\partial_k\sigma$ is the observed state using the communication channels and superposition.

If $\sigma \in \mathrm{Map}_k(f)$, it follows from the definition that $\partial\sigma \in \mathrm{Map}_{k-1}(f)$, and therefore we obtain the sequence of linear maps

$$\cdots 0 \to \mathrm{Map}_L(f) \xrightarrow{\partial_L} \mathrm{Map}_{L-1}(f) \xrightarrow{\partial_{L-1}} \cdots \xrightarrow{\partial_1} \mathrm{Map}_0(f) \to 0 \cdots . \tag{56}$$

We note that (56) is akin to Diagram (15). What we need to show is that there is no pass-through information. This follows from the fact that the maps ∂_k have the important property

$$\partial_{k-1} \circ \partial_k = 0. \tag{57}$$

This is a consequence of the assumption that any rigid broken trajectory is connected to a different rigid broken trajectory by a one-dimensional family of flow lines; recall the Figures 4, 5 and 6, and the discussion in Chapter 5.

Here is the argument for (57) using the "assumed properties" listed in 18.

(58)

$$\begin{aligned}
(\partial_{k-1}\partial_k\sigma)(x) &= \sum_{\{y\ |m(y)=m(x)+1\}} \chi(x,y)(\partial\sigma)(y) \\
&= \sum_{\{y\ |m(y)=m(x)+1\}} \chi(x,y) \sum_{\{z\ |\ m(z)=m(y)+1\}} \chi(y,z)\cdot\sigma(z) \\
&= \sum_{\{z\ |\ m(z)=m(x)+2\}} \left(\sum_{\{y\ |\ m(y)=m(x)+1\}} \chi(x,y)\cdot\chi(y,z) \right) \cdot \sigma(z). \\
&= \sum_{\{z\ |\ m(z)=m(x)+2\}} \boldsymbol{\chi}(x,z)\cdot\sigma(z).
\end{aligned}$$

As already indicated, $\boldsymbol{\chi}(x,z)$ for $m(z)=m(x)+2$ is defined by

$$\boldsymbol{\chi}(x,z) = \sum_{\{y\ |\ m(y)=m(x)+1\}} \chi(x,y)\cdot\chi(y,z). \tag{59}$$

The number $\boldsymbol{\chi}(x,z) \in \mathbb{Z}_2$ counts once broken rigid flow lines going from z to x, where the count is 0 if we have an even number and the count is 1 if we have an odd number. But, since by assumption the broken trajectories from x to z occur in pairs, their number is even. This means that we always have

$$\boldsymbol{\chi}(x,z) = 0. \tag{60}$$

This implies the following fundamental property.

Proposition. For every $\sigma \in \mathrm{Map}_k(f)$ the following identity holds

$$\partial_{k-1}(\partial_k(\sigma)) = 0.$$

Usually, mathematicians suppress the index k, and one can abbreviate the above statement as $\partial \circ \partial = 0$. This is an instance of the structure Henri Cartan was marvelling at when he received the Doctor Honoris Causa from Oxford; see page 53.

Taking from (56) the segment

$$\mathrm{Map}_{L+1}(f) \xrightarrow{\partial_{L+1}} \mathrm{Map}_L(f) \xrightarrow{\partial_L} \mathrm{Map}_{L-1}(f),$$

we can identify the collection of stealthy states on level L with the kernel $\ker(\partial_L)$ of ∂_L. The fragments coming from level $L+1$ constitute the image $\mathrm{im}(\partial_{L+1})$ of ∂_{L+1}, and finally, the unspoiled proprietary information on level L is the quotient vector space $\ker(\partial_L)/\mathrm{im}(\partial_{L+1})$.

Thus, we have constructed what mathematicians call a chain complex; in our case $(\mathrm{Map}(f), \partial)$. Chain complexes are studied within the area of homological algebra and are abundant in many areas of mathematics. In our context, they store certain algebraic information about $f : \Lambda \to \mathbb{R}$ by analyzing the negative gradient flow via its rigid flow lines and rest points. The next discussion will produce an algebraic procedure which, from $(\mathrm{Map}(f), \partial)$, extracts information only depending on Λ and not on f.

I.3 Removing Noise. Using (56), denote by $\ker_k(f)$ the subset of all maps in $\mathrm{Map}_k(f)$ with the property that $\partial\sigma = 0$. In our picturesque description, these are all states involving only lights on the critical points of Morse index k, with the property that our communication channels and superposition do not distinguish them downstream on level $k-1$ from the state where all lights are switched off.

Define the collection $\mathrm{im}_{k+1}(f)$ to consist of all $\partial\sigma$, where σ varies over $\mathrm{Map}_{k+1}(f)$. We can view this set as the collection of all possible observations on level k—via the channels and using superposition—coming from states involving critical points of Morse index $k+1$. Since $\partial \circ \partial = 0$, it follows that $\mathrm{im}_{k+1}(f)$ is a linear subspace of $\ker_k(f)$. Hence, we have among the states associated to the critical points with Morse index k, which cannot be distinguished via ∂ from O on level $k-1$, those states on level k which are obtained from observations from states on level $k+1$. Since $\mathrm{im}_{k+1}(f) \subset \ker_k(f)$, linear algebra tells us that we can take the quotient vector space $\ker_k(f)/\mathrm{im}_{k+1}(f)$. This quotient can be viewed as the proprietary data on level k discussed on page 54; see in particular (17) and Figure 2.

Whether this is a good idea for our original purpose is not yet clear—but it will indeed turn out to be an excellent idea. Indeed, we can marvel with Cartan about another instance of the concise proposition $d^2 = 0$.

More precisely, we can organize the states $\sigma \in \mathrm{Map}_k(f)$, which cannot be distinguished from O by ∂, into different groups by defining the sets

$$[\sigma] := \{\sigma + \tau \mid \tau \in \mathrm{im}_{k+1}(f)\},$$

where $\sigma \in \ker_k(f)$. Any two such sets are either the same or disjoint, i.e.; $[\sigma] = [\sigma']$ or $[\sigma] \cap [\sigma'] = \emptyset$. Indeed, $[\sigma] = [\sigma']$ if and only if $\sigma - \sigma' \in \mathrm{im}_{k+1}(f)$. Moreover, $[O] = \mathrm{im}_{k+1}(f)$. All the elements in $[\sigma]$ cannot be distinguished from O by ∂, but given two elements $\tau, \tau' \in [\sigma]$ their difference state $\tau - \tau'$ is a possible observation coming via ∂ from the critical points with Morse indicex equal to $k+1$.

From a mathematical perspective, the collection of all $[\sigma]$ defines the elements of the (quotient) vector space

(61) $$H_k(f) := \ker_k(f)/\mathrm{im}_{k+1}(f).$$

The $\mathbb{Z}_2$-vector space $H_k(f)$ is called the k-th homology group with coefficients in $\mathbb{Z}_2$ associated with the Morse function f.[152]

The homology $H(f)$ of f is the graded vector space

$$H_*(f) = \bigoplus_{k=0}^{L} H_k(f) = H_0(f) \oplus H_1(f) \oplus \ldots \oplus H_L(f).$$

Since

$$\mathrm{im}_{k+1}(f) \subset \ker_k(f) \subset \mathrm{Map}_k(f),$$

it follows that for every $k \in \mathbb{Z}$

$$\sharp\,\mathrm{Crit}_k(f) = \dim \mathrm{Map}_k(f) \geq \dim \ker_k(f) \geq \dim H_k(f).$$

This means that we have at least as many critical points of Morse index k as the dimension of the vector space $H_k(f)$.

The big result about $(\mathrm{Map}(f), \partial)$, not given here, is that for a suitable class of smooth spaces Λ, for every generic $f, g : \Lambda \to \mathbb{R}$, it holds that $H_k(f)$ is naturally(!) isomorphic to $H_k(g)$, and in particular they have the same dimension. This big result is established by a sizable mathematical construction, not described here, which associates to a pair (f, h) of Morse functions on Λ a linear vector space isomorphism $\Gamma_{(f,h)} : H_*(f) \to H_*(h)$ respecting the grading, i.e., mapping $H_k(f)$ to $H_k(h)$. This construction has the property that for three Morse functions (g, h, f), it holds

$$\Gamma_{(f,g)} = \Gamma_{(h,g)} \circ \Gamma_{(f,h)},$$

which just means that there are identifications (of their elements) $H_k(f) \approx H_k(h)$, $H_k(h) \approx H_k(g)$ and $H_k(g) \approx H_k(f)$ which are completely compatible.

The last statement means the following: For the smooth space Λ fixing a Morse function (and a generic Riemannian metric, a fact we shall suppress) we can construct $H(f)$. If we now take different data of the same kind, we not only obtain $H(g)$, but the whole procedure will also give us a precise method how to relate a specific element in $H_k(f)$ to a specific element in $H_k(g)$ by a linear isomorphism.

In other words, up to natural identifications, $H_k(f)$ has nothing to do with f, but is an invariant of Λ, which we denote by $H_k(\Lambda)$, called the k-th homology group of Λ with coefficients in $\mathbb{Z}_2$. The proof of this fact is formidable and can be found in [**Sch**]. Hence

$$\sharp\,\mathrm{Crit}_k > \dim H_k(\Lambda) \quad \text{for all } k \in \mathbb{Z}.$$

The relationship with previous discussions is that, with

$$P_\Lambda(t) = \sum_k \dim_{\mathbb{Z}_2} H_k(\Lambda) \cdot t^k$$

and

$$P_f(t) = \sum_k \sharp\,\mathrm{Crit}_k(f) \cdot t^k$$

one can show that these polynomials are related in the same way we already discussed before:

(62) $$P_f(t) = P_\Lambda(t) + (1+t) \cdot Q_f(t),$$

where the coefficients of Q_f are all non-negative.

While the Morse equality is one way to describe $f : \Lambda \to \mathbb{R}$, we just have seen another way to describe f (and a generic metric), namely the chain complex $(\mathrm{Map}(f), \partial)$. As we have mentioned in Subsection 1.1.2, there are in general many ways of describing a mathematical object. In our case, with the main phenomenon being the birth-death mechanism, we have both the polynomial description as well as the description via chain complexes, the latter not only implying the former, but also being more concise.

In summary, Witten's construction allows us to define the homology of the underlying smooth space from the combinatorics of a graph capturing the structure of rigid flow lines associated to the gradient flow of a given Morse function, i.e.

$$\text{Gradient Flows} \overset{\text{Morse-Smale genericity}}{\Longrightarrow} \text{Homological Algebra}$$

Philosophically, the same approach applied to a system of suitable partial differential equations will result in Floer homology. However, the number of mathematical issues which need to be surmounted increases dramatically.

$$\text{Systems of PDEs} \overset{\text{Lots of Analysis}}{\Longrightarrow} \text{Homological Algebra}$$

APPENDIX J

From Cartoon to Theory

We have described in Section 4.3 a corollary of Witten's approach to Morse theory: One can recover the homology of a compact smooth manifold by knowing the critical points and counting the rigid flow lines between them.[153] Floer's work extended this idea to an infinite-dimensional context by studying a Lagrangian intersection problem. Here we shall describe the easier problem of symplectic fixed points—the issues are similar.

In order to avoid serious analytical and transversality issues, Floer imposed topological assumptions; therefore, he did not prove the Arnold conjecture in the most general cases. His work was a quantum leap as far as the Arnold conjecture was concerned, but more importantly it introduced a powerful way to study—in various areas of mathematics—the types of problems we've been discussing. In the intervening years, with the same underlying idea, the methods have been extended and are now applicable to an amazing array of new types of problems.

The General Problem and the Associated Difficulties

We start with a compact symplectic manifold (M, ω) and a smooth Hamiltonian $H : S^1 \times M \to \mathbb{R}$, which defines a time-dependent Hamiltonian vector field X_{H_t}, $t \in S^1$. The goal is to establish an algebraic count of the 1-periodic orbits of the problem

$$\dot{z}(t) = X_{H_t}(z(t)), \quad t \in S^1, \tag{63}$$

which is independent of the choice of a generic H. One can show without too much effort the following:

(A1) If the C^1-norm of H is small enough, all periodic orbits are necessarily contractible.

(A2) If H is autonomous, i.e. does not depend on t, and a Morse function with a sufficiently small C^2-norm, then the 1-periodic solutions to the associated Hamiltonian system are precisely the critical points of H viewed as 1-periodic loops.

Establishing the assertion (A1) is straightforward. The argument for (A2), which is a bit trickier but well-known, can be found in [**HZ2**], Chapter 5, Proposition 17.

In view of (A1), any reasonable algebraic count of non-contractible 1-periodic orbits which does not depend on the choice of H should be 0.

Point (A2) will be useful when making computations of the invariants that Floer introduces. In fact, there will be a connection to the Witten complex for $-H$.

In view of the above, Floer picked the following strategy:

(S1) Define for sufficiently generic H a Witten-type complex—now called Floer complex or Floer-Witten complex— where the critical points are 1-periodic contractible orbits.

(S2) Show that for different Hamiltonians H and K the Floer complexes, are in some sense, isomorphic in some homological algebra sense.

(S3) Show that H satisfying (A2) is generic enough so that the Floer complex for the Hamiltonian system—as in (S1)—is defined and that the usual Morse theoretic Witten complex is defined for $-H$. Furthermore, show that both are isomorphic.

If these steps can be carried out, we have an algebraic count of contractible 1-periodic orbits of sufficiently generic Hamiltonian systems which is directly related to Morse theoretic counting of critical points via homological algebra. Such a relationship obviously addresses Arnold's conjecture, even if the results for certain symplectic manifolds are weaker than what Arnold predicted. The reason for this is that, possibly for certain symplectic manifolds, the homological lower bounds for critical points of a Morse function can be smaller than the minimal number of critical points a Morse function needs to have—a somewhat esoteric problem in our context. In the case of symplectic tori and many other manifolds, however, the method confirms Arnold's prediction.

Difficulties. As we already mentioned, quite a number of technical issues arise, which forced Floer to make specific assumptions on the topological properties of (M, ω).[154]

For instance, some of the difficulties that appear in general cases include:

(P1) The problem (63) is, in general, only locally variational, but not globally variational. It still makes sense to talk about critical points of locally variational problems. This phenomenon, in the context of Morse theory, can be illustrated by the following examples for the smooth manifold $S^1 = \mathbb{R}/\mathbb{Z}$. The variational case: A smooth Morse function on S^1 has at least two critical points. The locally variational case: The 1-form dt can be locally written as the (exterior) derivative of a smooth function of the form $t \to a + t$, where a is a constant. In this example, we see that the number of critical points are different. Namely, in the Morse case there are always at least two critical points and in the locally variational case there might not be any.

(P2) The attempt at using the Conley-Zehnder index μ_{CZ} as the substitute for the infinite Morse index, which works for T^{2n}, cannot be carried out in general. Instead, one has to live with $\mu_{CZ} \bmod (2N)$, the reduced Conley-Zehnder index, where $N \geq 0$ is an integer depending on the topology of (M, ω). If $N = 0$, we are in a similar situation as for the torus, but that is not the case if $N > 0$. For example, if $N = 1$, one can only say that the Conley-Zehnder index is even or odd. In other words, in general, the right notion leads to cyclic numbers $\mathbb{Z}/(2N\mathbb{Z}) = \{0, 1, \ldots, 2N-1\}$, i.e. $2N \equiv 0$—a new phenomenon, since the difference of the infinite Morse indices makes only sense as cyclic numbers. This is related to the fact that if $\pi_2(M) \neq 0$, the smooth loop space $C^\infty_{contr}(S^1, M)$ has nontrivial free homotopy classes.[155]

(P3) A version of Uhlenbeck's bubbling-off phenomenon arises, which causes analytical difficulties with respect to compactness and transversality.

What was described in (P2) is an infinite-dimensional phenomenon. In the torus case, we highlighted the fact that one can talk about the difference of two infinite Morse indices and explained it as the difference of two (integer) Conley-Zehnder indices. For other compact symplectic manifolds, it turns out that there can be an ambiguity modulo $2N$ for the difference, where N is a nonnegative integer. For the simple example of the two-sphere $\mathbb{S}^2$, the ambiguity is an integer multiple of 4, so $N = 2$.

The underlying reason is that, in order to define the Conley-Zehnder index for a (contractible) periodic orbit for a general symplectic manifold, we need to pick a map from the closed unit disk $\mathbb{D} \subset \mathbb{C}$, which, if restricted to the boundary, parametrizes the periodic orbit. The problem is that the definition of the Conley-Zehnder index might depend on the choice of the disk. In which way does it depend on disk choice? A careful analysis shows that the definition might only depend on the relative homotopy class of the disk. But is that indeed the case? It depends! A necessary condition for a nontrivial dependence on the homotopy class is the existence of a map $\mathbb{S}^2 \to M$ which is non-contractible.

One can also express the underlying reason in the following way: For two given contractible periodic orbits z_1 and z_2, which are in the same connected component of M, we can take a path in $C^\infty_{\mathrm{contr}}(S^1, M)$ connecting z_1 with z_2. Such a path is nothing else but a map

$$u : [0,1] \times S^1 \to M,$$

satisfying $u(0,t) = z_1(t)$ and $u(1,t) = z_2(t)$ for all $t \in S^1$. Given this path, we can define the "relative Conley-Zehnder index with respect to the path" as an integer without assigning Conley-Zehnder indices to z_1 or z_2. Unfortunately, generally this difference depends on the choice of the homotopy class of the connecting path. Taking a different path, the results of the computation of the relative Conley-Zehnder indices will differ by a multiple of $2N$. This fact is not immediately obvious, and it takes some work to establish. Hence, if we want to define a relative Conley-Zehnder index that is independent of the path, we have to take the computation for any chosen path and mod out by $2N\mathbb{Z}$. Having now a mod $2N$ relative Conley-Zehnder index, we can write it as the difference of two mod $2N$ indices—that is, the Conley-Zehnder indices mod $2N$, assigned to the periodic orbits.

It turns out that if difficulties (P1) and (P2) do not arise, (P3) will also not arise. That this is the case follows from a lengthy analytical argument. In order to show some compactness property for bounded flow-lines of the unregularized gradient flow, one carries out, assuming there is no compactness, a version of Uhlenbeck's bubbling-off analysis—the result shows that a nontrivial holomorphic sphere exists. This turns out to be impossible if we assume that the problem is globally variational.

Floer's Choice—The Good Case

Given the mountain of difficulties described above, Floer made some additional assumptions. He assumed that the problem is globally variational and that the Conley-Zehnder index is defined as an integer. These assumptions prevent bubbling-off, but still require hard analytical work. At the end of a lot of work, a version of the cartoon pictures that we have seen in finite-dimensional Morse theory holds for a Witten-type construction for the unregularized gradient flow associated to $\mathcal{A}_H$.

The Global Functional. Floer assumes that the symplectic form ω vanishes over $\pi_2(M)$. This means that given any smooth enough map $u : \mathbb{S}^2 \to M$, we have the equality

$$\int_{\mathbb{S}^2} u^*\omega = 0. \tag{64}$$

This prompts the following definition.

DEFINITION. *If (M, ω) is a symplectic manifold and the condition (64) holds, we say that the symplectic class vanishes on the second homotopy group abbreviated as $[\omega]|\pi_2(M) = 0$.*

A priori, as follows from the theory of differential forms, the integral in (64) depends only on the homotopy class $[u]$ of u. Under Floer's assumptions on (M, ω), which we just described, we consider the space $C^\infty_{\text{contr}}(S^1, M)$ of contractible smooth loops. First, we define the action functional on this space. For a given $z \in C^\infty_{\text{contr}}(S^1, M)$, we find a smooth map u defined on the closed two disk $\mathbb{D}^2$ having the property

$$u(e^{2\pi i t}) = z(t) \text{ for all } t \in S^1.$$

The action is then defined by

$$\mathcal{A}(z) = \int_D u^*\omega.$$

Because of the assumptions, this number is independent of the particular choice of u. Fixing a Hamiltonian H we add a Hamiltonian term to the action to obtain

$$\mathcal{A}_H : C^\infty_{\text{contr}}(S^1, M) \to \mathbb{R} : z \to \mathcal{A}(z) - \int_{S^1} H(t, z(t)) \cdot dt. \tag{65}$$

At least formally the critical points of $\mathcal{A}_H$ are the contractible 1-periodic orbits of the Hamiltonian system associated to H. Of course, Floer theory is the "ticket" to study $\mathcal{A}_H$ on symplectic manifolds other than T^{2n}.

Generic Non-Degeneracy. As we have seen, in order to carry out a Morse theory in finite dimensions, we need the critical points to be non-degenerate. This is a condition on the Hessian. In the Floer case, a suitable replacement is the assumption that the linearized Hamiltonian system associated to a critical point only has the zero-solution as a 1-periodic orbit. This means the following: We use H to define, via the associated ordinary differential equation, the one-parameter arc of symplectic diffeomorphisms

$$[0, 1] \ni t \to \Phi_t, \quad \Phi_t : (M, \omega) \to (M, \omega),$$

where $\Phi_0 = \text{Id}$. Since z is a 1-periodic orbit, it holds $z(0) = z(1)$ and $\Phi_1(z(0)) = z(0)$. We take the linearization of Φ_1 at $z(0)$ to obtain the linear tangent map

$$T\Phi_1(z(0)) : T_{z(0)}M \to T_{z(0)}M.$$

The non-degeneracy condition means precisely that 1 does not belong to the spectrum $\sigma(T\Phi_1(z(0)))$. In other words, there is no nonzero vector $h \in T_{z(0)}M$ that satisfies $T\Phi_1(z(0))(h) = h$.

Not every H satisfies this condition for every contractible periodic orbit, but most H satisfy this property in the sense of Baire category. This can be shown by applying the Sard-Smale theorem, [**Sm2**], to a properly designed problem—with a little bit of work one can show that, in fact, by an arbitrarily small C^∞-perturbation this can be achieved.

Generic Finiteness of Periodic Orbits and Conley-Zehnder Index. Assuming that H has the non-degeneracy property, one can show that every periodic orbit is isolated in $C^\infty_{\text{contr}}(S^1, M)$. An Ascoli-Arzelà compactness type argument applied to $\dot{z}(t) = X_t(z(t))$ implies that the set of periodic orbits in $C^\infty_{\text{contr}}(S^1, M)$ is compact. Compactness and isolation implies that there are only a finite number of periodic orbits.

Next we would like to define, for the finitely many contractible 1-periodic orbits, the Conley-Zehnder index. In case one wants to obtain an integer-valued Conley-Zehnder index, one needs to impose a topological assumption similar to $[\omega]|\pi_2(M)$, namely the condition that

$$c_1|\pi_2(M) = 0. \tag{66}$$

Here c_1 is called the first Chern class.

DEFINITION. *If for a symplectic manifold (M, ω) the property* (66) *holds, we shall say that the first Chern class vanishes on the second homotopy group of M.*

Chern classes are important invariants from algebraic topology associated to complex vector bundles over topological spaces, see [**MaT, BT**]. Without going into too much detail, in our case the first Chern class is an algebraic-topological measure of the twisting of the tangent bundle of the symplectic manifold (M, ω), using the fact that it can be viewed as a complex vector bundle.[156]

Assume we are given a compact symplectic manifold (M, ω) satisfying the two conditions:

$$[\omega]|\pi_2(M) = 0 \quad \text{and} \quad c_1|\pi_2(M) = 0.$$

With $H : S^1 \times M \to \mathbb{R}$ being a non-degenerate periodically forced Hamiltonian, we obtain a finite number of contractible 1-periodic orbits, say

$$z_1,\ z_2,\ \ldots,\ z_\ell$$

and think of them as shown in Figure 1.

One should keep in mind, however, that each periodic orbit z_i comes with additional data, the Conley-Zehnder index μ_i and the action $a_i := \mathcal{A}_H(z_i)$:

$$z_1 \equiv (z_1, \mu_1, a_1), \ldots,\ z_\ell \equiv (z_\ell, \mu_\ell, a_\ell).$$

Of course, at this point we do not even know if we actually have any periodic orbit—and it is part of Floer's theory to show that there are at least as many as Arnold predicted.

Adapted Almost Complex Structures and Riemannian Metrics. In order to write down an L^2-gradient for the action $\mathcal{A}_H$ defined in (65), we fix in a smooth way for every tangent space T_mM a linear map $J(m) : T_mM \to T_mM$ with the property that $J(m)^2 = -\text{Id}$, thus defining a smooth almost complex structure denoted by J. We require from J that $g_J := \omega \circ (\text{Id} \oplus J)$ is a Riemannian metric. As a consequence

$$\omega \circ (J \oplus J) = \omega.$$

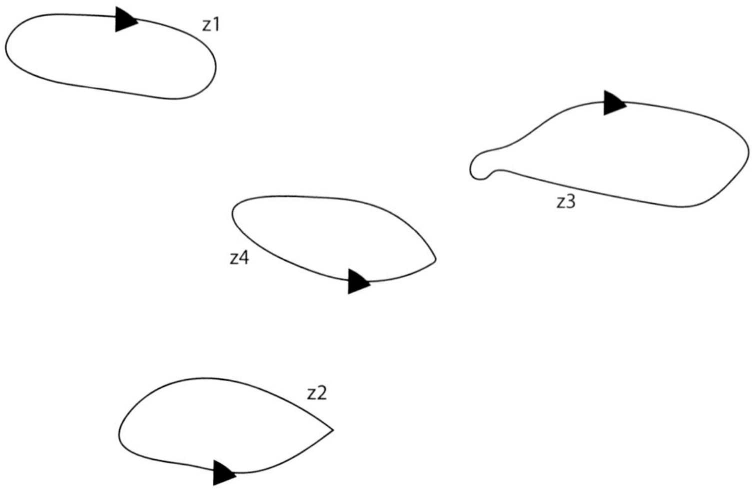

FIGURE 1. The non-degenerate periodic orbits; the flow-lines of the unregularized gradient flow will be cylinders connecting them.

Such a J is called a compatible (with ω) almost complex structure. It equips every tangent fiber $T_m M$ with the structure of a complex vector space, where multiplication by the complex number $\alpha + i\beta$ is defined by

$$(\alpha + i\beta) \cdot v := \alpha \cdot v + \beta \cdot J(m)v.$$

Gromov showed, [**G**], that such compatible almost complex structures J are very important. Associated to J we have the generalized Cauchy-Riemann type operator $\bar{\partial}_J$ mentioned in (35).

The L^2-Gradient and the Unregularized Gradient Flow. The derivative of $\mathcal{A}_H$ at z in the direction h is given by

$$d\mathcal{A}_H(z)(h) = \int_{S^1} g_J(z(t))(-J(z(t))\dot{z}(t) - \nabla_{g_J} H_t(z(t)), h(t)) \cdot dt. \tag{67}$$

Here $h \in L^2(z^*TM)$, i.e. h is a section of the pull-back bundle $z^*TM \to S^1$. Hence the L^2-gradient (associated to g_J) is given by

$$\nabla_{L^2}\mathcal{A}_H(z) = -J(z)\dot{z} - \nabla_{g_J} H_t(z)$$

and the unregularized gradient flow equation $\frac{\partial u}{\partial s} = \nabla_{L^2}\mathcal{A}_H(u)$, rewritten slightly, is

$$\bar{\partial}_J(u) + \nabla_{g_J} H_t(z) = 0. \tag{68}$$

Here, $u : \mathbb{R} \times S^1 \to M$ and

$$\bar{\partial}_J(u) = u_s + J(u)u_t. \tag{69}$$

We shall see that we cannot work with an arbitrary compatible J in order to address some of the fine points described later. However, a number of properties are already true without picking special J.

Assuming the non-degeneracy condition, one can show that a solution $u : \mathbb{R} \times S^1 \to M$ of (68), which also satisfies

$$-\infty < \inf_{s \in \mathbb{R}} \mathcal{A}_H(u(s)) \leq \sup_{s \in \mathbb{R}} \mathcal{A}_H(u(s)) < +\infty, \tag{70}$$

will converge exponentially fast as $s \to \pm\infty$ to periodic orbits, as in Figure 2.

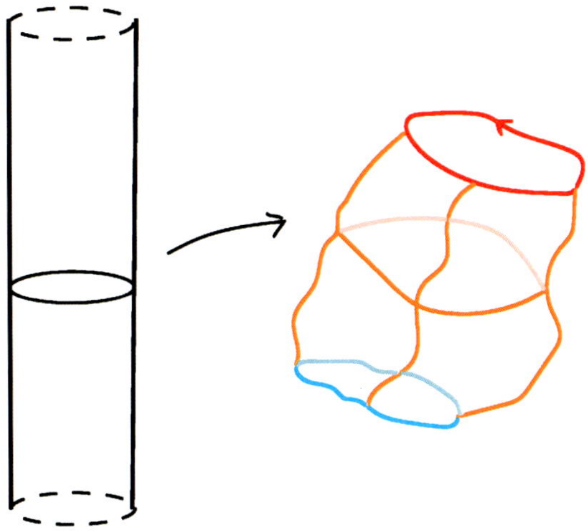

FIGURE 2. Two non-degenerate periodic orbits connected by one flow-line of the unregularized gradient flow. Geometrically it will be a map u from the standard infinite cylinder $\mathbb{R} \times S^1$ into M which satisfies (68) and (70)).

Further Choices and Genericity. If we pick two periodic orbits, we would like to know something about the structure of all bounded gradient-lines of the unregularized gradient flow connecting these orbits. It turns out that the equation (68) viewed in a suitable nonlinear analysis setup[157] will define a Fredholm section of a Hilbert space bundle over a Hilbert manifold. The linearization at a solution then defines a linear Fredholm operator. If this operator is surjective, we can conclude that the collection of all the gradient lines between the two fixed periodic orbits defines a manifold, its dimension being precisely the difference of the Conley-Zehnder indices. This surjectivity assumption is similar to the Morse-Smale assumption (3) in the "assumed properties" in 18. This is again a genericity question and one can achieve this genericity by either taking a suitable C^∞-small perturbation of H; or alternatively one can replace J by a time-dependent family J_t, $t \in S^1$, to achieve the same goal, with J_t "barely" depending on $t \in S^1$.

These perturbation arguments are very technical and subtle. Various issues, clarifying some of the ideas in Floer's seminal work and beyond, had been discussed by Floer and Hofer, as well as Hofer and Salamon, producing the posthumous paper [**FHS**].

Replacing J by a small perturbation of time-dependent compatible almost complex structures, say $S^1 \ni t \to J_t$ one can define the L^2-inner product for h, k sections of z^*TM by

$$(h,k)_z = \int_{S^1} g_{J_t}(z(t))(h(t),k(t)) \cdot dt.$$

In this case the unregularized flow-equation is

$$\frac{\partial u}{\partial s}(s,t) + J_t(u(s,t))\frac{\partial u}{\partial t}(s,t) + \nabla_{g_{J_t}} H(t,u(s,t)) = 0.$$

Using the Sard-Smale theorem [**Sm2**], which is a Fredholm version of the classical Sard theorem, [**Sa**], applied to a suitable set-up, one can show that for a generic loop $t \to J_t$ arbitrarily close to the original constant J, the linearized Fredholm sections are surjective at the solutions connecting to periodic orbits. As a consequence one obtains the following: Since $\mathbb{R}$ acts on the domain $\mathbb{R}\times S^1$ of the solutions, we see that if u is a solution so is u_τ defined by $u_\tau(s,t) = u(s+\tau,t)$. If the two critical points are different, then a nonempty solution space contains at least a one-dimensional manifold. Consequently, a solution space of dimension 1—the difference of Conley-Zehnder indices is 1—consists of rigid flow lines and the $\mathbb{R}$-action just gives their reparametrizations. The genericity discussed above corresponds to the requirement in finite-dimensional Morse theory that the Morse index difference is 1 and that the flow is Morse-Smale.

Compactness. Taking all the "generic precautions," one can show that there are only a finite number of 1-periodic orbits and that they are all non-degenerate. Further, given two periodic orbits of Conley-Zehnder index difference 1 the set of flow lines connecting the orbits is finite and the flow lines are rigid. Moreover, if the Conley-Zehnder index difference is 2, after dividing out by the $\mathbb{R}$-action, we obtain a one-dimensional family of flow lines. This family either has the structure of a circle parametrizing cylinders, or the family can be parametrized by the interval $(0,1)$. In the latter case, each $\tau \in (0,1)$ corresponds to a cylindrical flow line varying smoothly as a function of τ. As τ gets closer to 0 or 1, the flow lines decompose into rigid broken cylinders, a phenomenon similar to what we discussed in finite-dimensional Morse theory in Section 4.3, see Figure 4. The only difference is that lines are replaced by cylinders, see Figure 2 and Figure 3.

The generic choices of H and J will also guarantee that a one-dimensional family of flow lines emanates from a broken rigid flow line. The other end of the one-dimensional family splits up at a different broken configuration. This follows from a version of Taubes' gluing construction. The work necessary to make this mathematically sound is also highly technical compared to our cartoon versions.

The Floer Complex

At this point everything is in place to associate to (H,J) a chain complex similar to what we did in finite-dimensional Morse theory.

Our rigid gradient lines are solutions of

$$\frac{\partial u}{\partial s}(s) = \nabla_{L^2}\mathcal{A}_H(u(s)) \text{ and } \mathbb{R} \ni s \to \Phi_H(u(s)) \text{ is bounded.} \tag{71}$$

General Data. Floer's procedure will produce a cohomology group for (H,J). In certain cases it can be identified with the usual homology for M with $\mathbb{Z}_2$-coefficients. This precise identification between the cohomology group and homology group is somewhat subtle due to the fact that their gradings have to be adjusted—the correspondence of the gradings for the Floer groups and the Morse

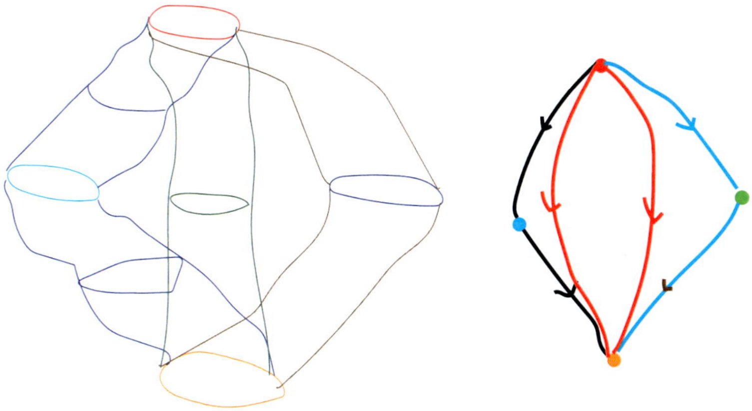

FIGURE 3. At left, two non-degenerate periodic orbits of Conley-Zehnder index difference 2. A one-dimensional family of cylinders connects two broken cylindrical configurations. At right, the equivalent picture in finite-dimensional Morse theory.

homology groups depends on the normalization of the Conley-Zehnder index. Since the Floer groups are generated by contractible periodic orbits, we obtain an estimate of their number in terms of the Morse homology—that is, the homology of the Witten complex. Since Morse homology is isomorphic to singular homology, this implies that the number of 1-periodic orbits can be estimated in terms of the "hole count" of M measured by singular homology theory with $\mathbb{Z}_2$-coefficients, see the top of page 16.

Before we venture into the thick of it, let us consider the construction's overall structure.

There are choices to be made: For one, we must pick a pair (H, J) with suitable properties. The main piece of data is H since we would like to say something about its 1-periodic orbits. However, the whole construction depends explicitly on J. Of course, we would like to end up with a result similar to

$$P_f(t) = P_\Lambda(t) + (1+t) \cdot Q_f(t), \tag{72}$$

echoing our discussion of Morse theory. Namely, there should be a count P_H of the number of periodic orbits for a specific H, a count $P_{(M,\omega)}$ of the "minimal" number of 1-periodic orbits that have to exist on (M, ω), and birth-death count Q_H resulting in a formula of the following kind

$$P_H(t) = P_{(M,\omega)}(t) + (1+t) \cdot Q_H(t). \tag{73}$$

We already have seen that the right language for achieving this is homological algebra, particularly the notion of chain complex (or cochain complexes) and their associated homology (or cohomology).[158] Here we shall consider only chain or cochain complexes which are also vector spaces over $\mathbb{Z}_2$.

The (Co-)Chain Complexes. We fix generic data $\mathfrak{c} = (H, J)$, with J possibly times-dependent, and with all the properties we have described before:

(1) The 1-periodic orbits associated to H are non-degenerate.
(2) The solutions $u : \mathbb{R} \times S^1 \to M$ of of $\bar{\partial}_J(u) + \nabla_{g_J} H_t(u) = 0$ that are Φ_H-bounded are described by nonlinear Fredholm operators. They are regular

in the sense that the linearizations are surjective. This is the equivalent of the Morse-Smale condition.

One can show, as a consequence of the non-degeneracy condition, that at the ends a solution u converges exponentially fast to periodic orbits.[159] We shall write u^+ and u^- for the asymptotic limits obtained from u as $s \to \pm\infty$. The rigid flow lines have the property that

$$\mu_{CZ}(u^+) - \mu_{CZ}(u^-) = 1.$$

Denote by $\mathcal{P}_H$ the finite collection of all critical points. The Conley-Zehnder index defines in the present case a map $\mu_{CZ} : \mathcal{P} \to \mathbb{Z}$ and we can decompose the set into a disjoint union

$$\mathcal{P}_H = \bigsqcup_{k \in \mathbb{Z}} \mathcal{P}_H^k,$$

where

$$\mathcal{P}_H^k = \{z \in \mathcal{P}_H \mid \mu_{CZ}(z) = k\}.$$

Now we continue in a familiar way—as described when we dealt with the Witten complex. Define $C^k(\mathfrak{c})$ to be the $\mathbb{Z}_2$-vector space of all maps

$$\mathcal{P}_H^k \to \mathbb{Z}_2.$$

An element in $C^k(\mathfrak{c})$ can be viewed as a formal linear combination of periodic orbits with Conley-Zehnder index k with coefficients in $\mathbb{Z}_2$. Using our light analogy: if the coefficient of a periodic orbit is 0, the light installed at the periodic orbit is off; otherwise it is on.

We write the index in $C^k(\mathfrak{c})$ as superscript since we will be dealing with cohomology, rather than homology. In homology, the linear maps ∂ between chain groups decreased the degree by 1, and in our situation the degree will increase by 1. In homological algebra or algebraic topology, cohomology often occurs as the construction dual to homology.[160]

Next we define a linear map $\delta : C^k(\mathfrak{c}) \to C^{k+1}(\mathfrak{c})$ which depends on a count of solutions $u : \mathbb{R} \times S^1 \to M$ of

$$(\bar{\partial}_J u)(s,t) + \nabla_{g_J} H(t, u(s,t)) = 0. \tag{74}$$

Note that J might have been chosen time-dependent to achieve transversality, and consequently the equation depends on $\mathfrak{c} = (H, J)$. In order to define δ we only have to explain its value on a 1-periodic orbit with $\mu_{CZ}(x) = k$.

Analogously, as for the Witten complex, we define for a 1-periodic orbit x with $\mu_{CZ}(x) = 1$ the $\delta(x)$ by

$$\delta(x) = \sum_{\{y \in \mathcal{P}_H^{k+1}\}} \chi(x,y) \cdot y,$$

where $\chi(x,y) \in \mathbb{Z}_2$ is the mod 2 count of rigid cylinders going from x to y, where $\mu_{CZ}(y) - \mu_{CZ}(x) = 1$. The rigid cylinders are defined by a partial differential equation using the data (H, J) so that $\delta = \delta_{(H,J)}$.

Our goal is the study of the dynamical system associated to H. Since for our constructions we need to pick some auxiliary data J, there is the worry that the end result will be affected by this choice. Indeed, on the level of cochain complexes this worry is justified, but passing to cohomology the "noise" introduced by the choices is magically cancelled by algebra—the cohomology does not depend on the choices. More precisely, for two choices $\mathfrak{c}$ and $\mathfrak{d}$ there is a canonical linear isomorphism

$$\psi^*_{\mathfrak{c}\mathfrak{d}} : H^*(\mathfrak{c}) \to H^*(\mathfrak{d}) \tag{75}$$

preserving the grading. This isomorphism is constructed by making additional choices to relate the cochain complexes associated to $\mathfrak{c}$ and $\mathfrak{d}$. This gives a cochain map

$$\psi_{\mathfrak{c}\mathfrak{d}} : C(\mathfrak{c}) \to C(\mathfrak{d}), \tag{76}$$

which might depend on the choices. However, it can be shown that the map induced in cohomology does not. Of course, all this has to be proved—quite a task.

The display (75), obtained from (76) by passing to cohomology, means that that for every integer k we have a linear isomorphism $\psi_{\mathfrak{c}\mathfrak{d};k} : H^k(\mathfrak{c}) \to H^k(\mathfrak{d})$. The important property one needs to show is that for three choices $\mathfrak{c}, \mathfrak{d}$ and $\mathfrak{e}$, the identity

$$\psi_{\mathfrak{c}\mathfrak{e}} = \psi_{\mathfrak{d}\mathfrak{e}} \circ \psi_{\mathfrak{c}\mathfrak{d}}$$

holds. So, if we make a construction for one choice of data, then the elements of the associated cohomology groups can be identified naturally with elements of the cohomology group associated to another choice. The arguments for the results discussed so far will depend on heavy use of the constructions in [**F1**].

At this point, we have a sequence of $\mathbb{Z}_2$-vector spaces that does not, up to natural identifications, depend on the choices we made. For that reason, we will suppress the choice data $\mathfrak{c}$ in the notation:

$$\ldots, H^i, \; H^{i+1}, \ldots, H^0, \ldots, H^j, \ldots$$

We have a new mathematical object H^* associated to a compact symplectic manifold (under some assumptions), but we do not know what it is.

By making a careful choice $\mathfrak{c}$, Floer showed that $H^*(\mathfrak{c})$ can been identified with the homology associated to the Witten complex of a Morse function f, which is accomplished in [**F3**].

Next we provide a little bit more details.

Special Autonomous Data. If H does not depend on $t \in S^1$, and J also does not depend on S^1, then one can show that if H is a small enough Morse function, the periodic orbits are critical points of H. The 1-periodic orbits are just constant periodic orbits $S^1 \ni t \to m$, where $m \in \text{Crit}(H)$. Moreover the solutions u of our partial differential equations do not depend explicitly on $t \in S^1$. In other words they have the form

$$u(s,t) = x(s).$$

And since u satisfies (71), and u has the special form, we see that x satisfies

$$\frac{dx}{ds}(s) = -\nabla_{g_J} H(x). \tag{77}$$

That is, x is a flow-line for the negative gradient flow. We also may assume, if J is carefully picked, that the negative gradient flow is Morse-Smale. This is an interesting statement, since it indicates that there is a relationship between the Morse theory on M and the Floer theory for this particular autonomous H. As it turns out this is correct and this relationship was used by Floer to compute his new (co-)homology theory. We should note that (71) has the form of a positive gradient flow and (77) that of a minus-gradient flow. If we use the conventions for the Witten complex we obtain a chain complex and a homology theory, whereas the conventions in (71) lead to a co-chain complex and a cohomology theory.[161] On the level of our discussion these differences are not too important.

Regarding the Relation of Different Complexes. We have not described the set-up for verifying that the algebraic data associated to different choices of auxiliary structure is related. In his original work [**F2**], Floer used a more complicated method for showing this fact, which he later abandoned. For more homotopy related questions in Floer homology, Michael Sullivan,[**Su**], showed that Floer's original approach provided more information, which sometimes can be useful. As long as one is only interested in Floer homology, the approach given in [**F13**] is easier, lying technically on the same level as showing that $\delta^2 = 0$ and we spare the reader the details. The main point, however, should be remebered: Given different choices of data $\mathfrak{c}$ and $\mathfrak{d}$, the elements of the corresponding cohomology groups can be naturally identified. Also, the associated cochain complexes are naturally quasi-isomorphic (a notion we shall not explain, but which allows the procedures in the next subsection).

Morse Relation for the Floer Complex. Again, as with the Witten complex, we have the property $\delta \circ \delta = 0$ and we obtain the co-chain complex $(C_H^*, \delta_{(H,J)})$. This cochain complex is constructed from all 1-periodic orbits, and notably the polynomial

$$P_H(t) = \sum_{k \in \mathbb{Z}} \dim_{\mathbb{Z}_2}(C^k(\mathfrak{c})) \cdot t^k$$

is quite similar to the polynomial P_f associated to a Morse function. Having a cochain complex we can pass to its cohomology

$$H^k(\mathfrak{c}) = \ker(\delta_{\mathfrak{c}}^k) / \operatorname{im}(\delta_{\mathfrak{c}}^{k-1}).$$

This again is a vector space over $\mathbb{Z}_2$ and we can define a polynomial

$$P_{(M,\omega)}(t) := \sum_{k \in \mathbb{Z}} \dim_{\mathbb{Z}_2}(H^k(\mathfrak{c})) \cdot t^k$$

corresponding to P_Λ. As we have seen $H^k(\mathfrak{c})$ is up to natural identification independent of $\mathfrak{c}$.

Since the C^k (for a fixed choice of data) are vector spaces over $\mathbb{Z}_2$, we can talk about their dimension and similarly for H^k. We decompose

$$C^k = \operatorname{im}(\delta_{k-1}) \oplus A^k \oplus B^k,$$

where $\operatorname{im}(\delta_{k-1}) \oplus A^k = \ker(\delta_k)$. Since $\delta_{k-1} : B^{k-1} \to \operatorname{im}(\delta_{k-1})$ is an isomorphism and $\dim(H^k) = \dim(A^k)$, we obtain the equality

$$\dim(C^k) = \dim(H^k) + \dim(B^{k-1}) + \dim(B^k) \quad \text{for all } k \in \mathbb{Z}.$$

We compute

$$\begin{aligned}
P_H(t) &= \sum_k \dim(C^k)\cdot t^k \\
&= \sum_k \big(\dim(H^k)\cdot t^k + \dim(B^{k-1})\cdot t^k + \dim(B^k)\cdot t^k\big) \\
&= \sum_k \big(\dim(H^k)\cdot t^k + t\cdot \dim(B^{k-1})\cdot t^{k-1} + \dim(B^k)\cdot t^k\big) \\
&= \sum_k \dim(H^k)\cdot t^k + t\cdot \sum_k \dim(B^{k-1})\cdot t^{k-1} + \sum_k \dim(B^k)\cdot t^k \\
&= P_{(M,\omega)}(t) + t\cdot \sum_k \dim(B^k)\cdot t^k + \sum_k \dim(B^k)\cdot t^k \\
&= P_{(M,\omega)}(t) + (1+t)\sum_k \dim(B^k)\cdot t^k \\
&= P_{(M,\omega)}(t) + (1+t)\cdot Q_{\mathfrak{c}}(t).
\end{aligned}$$

This shows that the cochain complex $(C^*(\mathfrak{c}), \delta_{\mathfrak{c}})$ contains the information about the critical points of $\mathcal{A}_H$—in such a way that we can obtain, by algebraic procedures, a formula of the kind we have derived before for Morse functions, see (73). Floer also showed that $P_{(M,\omega)}(t)$, with the right grading conventions, is the Poincaré polynomial coming from the Morse theoretic Witten complex.

We have sketched the general plan that crystallized in Floer's head and led to Floer theory. His success was not only a startling breakthrough; it was transformative, shifting the symplectic ecosystem. The global question could now be studied. The book [**AuDa**] by Michèle Audin and Mihai Damian gives a comprehensive introduction to what we discussed from a bird's eye view. In [**HZ2**] and [**McDS**] parts of Floer theory is also discussed.

Notes

1. This compelling framing was adopted from an anonymous reviewer (later determined to be John Ewing).

2. It would be better to consider symplectic geometry as the geometry associated to the stationary action principle, which for historic reasons is also commonly referred to as the principle of least action. The principle of stationary action states that the actual path taken by a system between two states is the one for which the action is stationary (i.e., it can be a minimum, maximum, or saddle point). In many cases, and certain formulations the action is minimized, which is why often it is called the principle of least action. However, there are scenarios where the action can be stationary without being a minimum or maximum. Thus, while the principle of least action is a specific case of the broader stationary action principle, the latter encompasses all situations where the action is stationary, not just those where it is minimized.

3. In his delightful book "The Best of All Possible Worlds" [**Ek2**] Ivar Ekeland takes the reader on a historical journey, starting with Maupertuis' ideas of the least action, and following its further developments and ramifications in physics, biology, ethics, and philosophy. As Freeman Dyson wrote in the New York Review of Books: " Ekeland gives us a vivid picture of human history and destiny... Mathematics appears as a unifying principle for history."

4. See [**H2**], $1:05:42$.

5. See [**Ek2**] for a vivid presentation of this circle of ideas.

6. This is an infinite-dimensional analogue of the one-dimensional (calculus) situation where we have a map $f:\mathbb{R}\to\mathbb{R}$ and we are looking for points $x\in\mathbb{R}$ at which the derivative $f'(x)$ is zero.

7. Bott contributed heavily to Morse theory and its applications to a broad range of problems, see e.g. his survey [**Bo1**].

8. September 1988, Paris

9. Mathematicians would write $f: I\to\mathbb{R}$, where I is a closed interval of the form $I=\{x \mid a\leq x\leq b\}$ in the real numbers $\mathbb{R}$. Here I represents the green line, a is the left boundary point and b the right one, and drawing all points of the form $(x, f(x))$, where x varies in I will give the graph colored in blue.

10. In more precise mathematical language: the condition is that at a point x with vanishing derivative, i.e. $f'(x)=0$, the second derivative is nonzero, i.e $f''(x)\neq 0$. Under this condition x is a local minimum (valley) provided $f'(x)=0$ and $f''(x)>0$ and a local maximum (peak) provided $f'(x)=0$ and $f''(x)<0$. The significance of $f''(x)$ being non-zero at a critical point x, i.e. $f'(x)=0$, is that the qualitative behavior of f near x is unambiguous. Namely at x we either have a strict local minimum or a strict local maximum. Due to results of Sard, [**Sa**], a small perturbation of a given smooth function will always have the property that valleys and peaks are well formed. The work of Sard is important for the existence of the field called differential topology, see [**MH**].

11. Polynomials can be added and multiplied. For this, view a polynomial as a formal infinite sum $\sum_{k=0}^{\infty} a_k\cdot t^k$, where only a finite number of the a_k's are non-zero. The addition and multiplication of polynomials is defined as follows:

$$\left(\sum a_k\cdot t^k\right)+\left(\sum b_k\cdot t^k\right)=\sum (a_k+b_k)\cdot t^k$$

and

$$\left(\sum a_k \cdot t^k\right) \cdot \left(\sum b_k \cdot t^k\right) = \sum_{k=0}^{\infty} \left(\sum_{i+j=k} (a_i \cdot b_j) \right) \cdot t^k.$$

12. We start changing to a more mathematical language. Here we talk about about what we previously called a well-formed valley or well-formed peak, respectively. A strict local minimum has the property $f''(x) > 0$ and a strict local maximum $f''(x) < 0$.
13. It is important for this statement that the valleys and peaks are well-formed, or more mathematically $f''(x) \neq 0$ at critical points.
14. In our three cases we can take for V from left to right: $\{0\}$, $\{(x, 0) \mid x \in \mathbb{R}\}$ and $V = \mathbb{R}^2$.
15. By the highest point of the pass we mean the lowest height one needs to accept to cross a pass.
16. When we go to higher dimensions we still have formulae of this kind. In certain space dimensions—and on more complicated spaces, called manifolds—it is generally not true (or not known) if $P_{\min}(t)$ actually can be attained by a suitable P_f. This difficulty is usually connected with properties of the fundamental group. To explain what this precisely means is beyond the scope of this book.
17. To see this we shall write $Q(t) := Q_f(t) = q_0 + q_1 \cdot t + q_2 \cdot t^2$. Then

$$(1+t) \cdot Q(t) = q_0 + (q_0 + q_1) \cdot t + (q_1 + q_2) \cdot t^2 + q_2 \cdot t^3,$$

where we know that q_0, q_1 and q_2 are non-negative integers. We rewrite the polynomial identity (7) as the system of identities

$$\begin{aligned} \sharp\,\mathrm{Crit}_0(f) &= 1 + q_0 \\ \sharp\,\mathrm{Crit}_1(f) &= 2 + (q_0 + q_1) \\ \sharp\,\mathrm{Crit}_2(f) &= 1 + (q_1 + q_2) \\ 0 &= q_2 \end{aligned}$$

This implies $q_2 = 0$ (Hence we might have assumed from the beginning that $Q_f(t) = q_0 + q_1 \cdot t$), and using that $q_0, q_0 + q_1, q_1$ are all non-negative that

$$\sharp\,\mathrm{Crit}_0(f) \geq 1, \ \sharp\,\mathrm{Crit}_1(f) \geq 2 \ \text{ and } \ \sharp\,\mathrm{Crit}_2(f) \geq 1$$

as claimed.
18. There are many more references to choose from, and what one ultimately picks is a matter of personal taste. For example [**AbM**] is a useful reference text if one thinks about dynamical systems on more general spaces.
19. A reader with some mathematical background can learn a lot more about mathematical holes in the nice Quanta article [**QC**]: The Hole Truth.
20. To be precise one needs to count holes of the pointed space $\Lambda/\emptyset$ obtained from the pair $(\Lambda, \emptyset)$ by collapsing $\emptyset$.
21. From the perspective of classical algebraic topology at the end of the seventies.
22. Physical laws are very often written as ordinary or partial differential equations. Kepler's laws are only approximations, but rather accurate since the sun is so much heavier than the planets and the gravitational force a planet experiences is essentially that coming from the sun and not the other planets.
23. See the book by Frauenfelder and van Koert, [**FvK**], the article [**FKM**] by Frauenfelder, Koh and Moreno as well as Dumas's book on KAM theory, [**Du**].
24. Analytic mechanics is an important step in the development of physics, for example it provides the context for Noether's theorem, which shaped particle physics.
25. It is only a recent development that chaos is being used in scientific space missions to save on fuel, see [**Be1, Be2**]. Edward Belbruno has been the leading figure in starting this development. Bringing anything from Earth into space is very expensive and fuel savings can mean significant payload increases. Of course, leaving the Earth's gravitational field will be always expensive, but very often one can save huge amounts of fuel when going into a nice orbit say around the Moon or Mars. These ideas were used for the NASA's GRAIL mission,

(Gravity Recovery And Interior Laboratory), which was launched in September 2011. Very recently also Floer theoretic tools have been used to built mathematical tools for finding suitable orbits,[**BFvK**, **FKM**].

26. If the goal is the study of the properties of a solution of an equation, one might think that the solution first must be computed—every mathematician before Poincaré thought so too. Poincaré reasoned that the equations which determine the solution might already provide nontrivial information about them. This idea turned out to be very powerful. It required completely new mathematical ideas and brought several mathematical fields into being.
27. The Poincaré-Birkhoff fixed point theorem is an example.
28. On some level symplectic geometry has been around for longer, but the developments which made it the field which it is today started in the early eighties.
29. See the article by S. Alan Stern in [**EnS**].
30. The Eddington number is the number of protons in the observable universe. In a lecture in 1938, Eddington proclaimed "I believe there are 15 747 724 136 275 002 577 605 653 961 181 555 468 044 717 914 527 116 709 366 231 425 076 185 631 031 296 protons in the universe and the same number of electrons." The current estimate is about six times this number, namely 10^{80}. On the cosmological length scale the universe is considered electrically neutral so that the 10^{80} protons correspond to about the same number of electrons. Therefore assuming that there are 10^{80} electrons, an upper bound on the theoretical size of a memory chip can be derived. Its theoretical maximal size limits the accuracy of the computations for Pluto to such an extent that they become meaningless beyond a time span of eighty million years due to accumulated mathematical errors from unavoidable round-offs and further processing of these numbers. In the next 80 million years Pluto might leave the solar system, crashes into the Sun, or crashes into the Earth—we just do not know. We do not have the resources to find out. Still we can compute that at least for a while we will be fine.
31. Alternatively we could also think of the map c as defined on the unit interval $[0,1]$ with the additional condition that $c(0) = c(1)$, i.e. it closes up.
32. This can be done by an approximation argument. Even more so one can define the action for $\frac{1}{2}$-times differentiable maps—what that means goes beyond the level of the book and the mathematics needed comes from interpolation theory, see Hans Triebel's book [**Tr**].
33. If we write
$$z(t) = (z_1(t), \dots, z_n(t)) = ((q_1(t), p_1(t)), \dots, (q_n(t), p_n(t)), \ t \in S^1 = \mathbb{R}/\mathbb{Z}$$
one can show that the value of $\mathcal{A}_{\mathbb{R}^{2n}}$ on a loop z is given by the following formula, where the last equality follows from a partial integration:
$$\mathcal{A}_{\mathbb{R}^{2n}}(z) = -\sum_{i=1}^{n} \int_0^1 p_i(t)\dot{q}_i(t)dt = \sum_{i=1}^{n} \int_0^1 q_i(t)\dot{p}_i(t)dt.$$
34. For example one is interested in orbits starting at earth, going several times nicely around the moon, and using substantially less fuel than the usual three day Hohmann transfer. A celebrated breakthrough solution taking 180 days was found by Belbruno in 1990 and started a whole development. For the Hohmann transfer 20% of the mass of the spacecraft arriving near the moon consists of fuel—it is needed to bring the spacecraft on a nice orbit around the moon. Belbruno's method basically does not need this fuel, but it takes 180 days. This was later improved to 90 days. The question is whether there are shorter time solutions which do not need fuel at the end. These new type of solutions, which use no or very little fuel at the end, make heavy use of the fact that the system describing the movement of a satellite in the sun-earth-moon system is chaotic. Rather than using lots of fuel one can use a combination of small amounts of fuel and the gravitational forces of moon-earth-sun to do the job.
35. The word "symplectic" was coined by Hermann Weyl [**Wey**], and is a calque of "complex." What is now the 'symplectic group' had been formerly called the 'line complex group.'
36. A property of a closed two-form is that it can be written locally as the exterior derivative of a one-from λ—every point $m \in M$ admits an open contractible neighborhood $U = U(m)$ on which a λ_U is defined satisfying $d\lambda_U = \omega$. The assignment of a number to a small loop $z : S^1 \to U(m)$ is then precisely the integration of $z^*\lambda$ over the circle S^1.

37. Species: Japanese Macaque (aka Snow Monkey), Macaca fuscata. Location: Jigokudani in the mountains of Japan, by the famous hot spring where wild monkeys come to bathe during the winter and seemingly have deep thoughts. Photo by Mogens Trolle.

38. Passing to what mathematicians call a suitable covering of the space of contractible loops one can define a single valued action functional. The possibly multi-valued map which we have mentioned is obtained as follows. For a map $Z : \mathbb{D} \to M$, where $\mathbb{D} = \{x + iy \mid x^2 + y^2 \leq 1\}$ is the closed unit disk in the complex plane $\mathbb{C}$ define

$$\mathcal{A}(Z) = \int_{\mathbb{D}} Z^*\omega = \int_{\mathbb{D}} \omega\left(\frac{\partial Z}{\partial x}, \frac{\partial Z}{\partial y}\right) dxdy.$$

Define two such disk maps Z and Z' to be equivalent provided
(1) $Z(x+iy) = Z'(x+iy)$ for all $x+iy$ with $x^2+y^2 = 1$, i.e. $Z = Z'$ on the boundary $\partial\mathbb{D}$.
(2) Z can be smoothly deformed into Z' by keeping the boundary values fixed.
We shall write $[Z]$ for the equivalence class associated to Z. One can show by an application of Stokes' Theorem for differential forms that for $Z' \in [Z]$ we have that $\mathcal{A}(Z) = \mathcal{A}(Z')$. We shall call $z = [Z]_{|\partial\mathbb{D}}$ the boundary value of $[Z]$ which is just the restriction of any representative of $[Z]$ to $\partial\mathbb{D}$. It can happen that for $[Z] \neq [Z']$ it is still true that $[Z]_{|\partial\mathbb{D}} = [Z']_{|\partial\mathbb{D}}$, but the values $\mathcal{A}(Z)$ and $\mathcal{A}(Z')$ are different. We denote by $\mathcal{A}([Z])$ the value of $\mathcal{A}(Z)$, which is independent of the choice of representative in the class $[Z]$. Finally, for a contractible loop $z : \mathbb{R}/\mathbb{Z} \to M$ we can define

$$\mathcal{A}(z) = \{\mathcal{A}([Z]) \mid [Z]_{|\partial\mathbb{D}} = z\}.$$

This is the possibly multi-valued functional. As an example consider the (oriented) two-sphere equipped with a volume form ω of total volume $V > 0$. We leave it to the reader to show that $\mathcal{A}(z) = \mathcal{A}(Z) + V \cdot \mathbb{Z}$, where $Z : \mathbb{D} \to S^2$ has boundary value z.

39. Removing the smallness assumption opens a can of worms. Suddenly the cohomological properties of ω and the first Chern class c_1 enter and one only obtains a variational problem on a suitable covering space of the loop space. Also the Conley-Zehnder indices, replacing the infinite Morse indices, behave as if they are cyclic numbers in some $\mathbb{Z}_{2N}$, for a suitable integer $N \geq 0$. There is no indication that Arnold was aware of special features and difficulties of his proposed generalization.

40. There is no shortage of lore about Pope Moser: Edward Belbruno, famous for introducing the use of chaos theory into space travel, [**Be1**, **Be2**], recalled the following story: "In 1993 I was in Zurich, on my way to Paris, and Jürgen Moser invited me to give a lecture at ETH in the mathematics seminar. It was on my recent work in celestial mechanics and how it was applied to space travel to get a Japanese spacecraft to the Moon. Jürgen was my doctoral advisor at NYU's Courant Institute. I was his last student there in 1981 before he relocated to ETH. I was excited to be sharing my new work. When we were standing in the dining room at ETH before having lunch he described to me an extraordinary and chilling story of how he made his way to the west at the end of World War II: It was early May 1945 and Germany had just surrendered. Jürgen was a soldier in the German army at the border between the Soviet and western zones in Germany. He desperately wanted to make his way to the west. The Soviet troops were overrunning the area he was in. He saw them coming in the distance. He said that he ran into a farm house to hide and into the basement. The basement was flooded. He was standing in the cold filthy water in the dark with the water up to his neck. He heard some Soviet troops go into the farm house and they were on the floor above him. He dared not move and stood there quietly in the dark. He said he stood there for three days while the soldiers occupied the building. As soon as he heard them leave, he went upstairs and quickly ran across the border to safety to western Germany.

41. Around the same time, the mathematics department at ETH was hiring. The committee agreed on a suitable candidate for whom letters of reference should be solicited. Moser was among the desired letter writers. The conscientious committee chair, Heinrich Ursprung, a developmental biologist and the president of the ETH, asked about Moser's scientific standing. When the committee members outlined his stature, the chair wondered why they were going after secondary figures rather than this exemplary scientist. The committee informed him that, alas, there was simply no chance of acquiring the pope. Undeterred, Ursprung—according to Zehnder an exemplary ETH President—simply called Moser. Moser started his

professorship at the ETH in 1981, and later served as director of the Forschungsinstitut für Mathematik (FIM).

42. In the context this means not suitable for global questions in contrast to local questions.

43. Written more explicitly with $S^1 = \mathbb{R}/\mathbb{Z}$ and $z = ((x_1, y_1), \ldots, (x_n, y_n))$

$$S(\lambda, z) = \int_{S^1} \left(\left(\sum_{k=1}^{n} y_k(t) \cdot \frac{dx_k}{dt}(t) \right) - \lambda \cdot (H(z(t)) - c) \right) \cdot dt.$$

One can show by writing $y_k \cdot \frac{dx_k}{dt} = \frac{1}{2} \cdot (y_k \cdot \frac{dx_k}{dt}) + \frac{1}{2} \cdot (y_k \cdot \frac{dx_k}{dt})$ and then applying to one of the terms on the right a partial integration that

$$\begin{aligned} S(\lambda, z) &= \int_{S^1} \left(\frac{1}{2} \cdot \sum_{k=1}^{n} \left(y_k(t) \cdot \frac{dx_k}{dt}(t) - x_k(t) \cdot \frac{dy_k}{dt}(t) \right) - \lambda \cdot (H(z(t)) - c) \right) \cdot dt \\ &= -\mathcal{A}_{\mathbb{R}^{2n}}(z) - \lambda \cdot \int_{S^1} (H(z(t)) - c) \cdot dt. \end{aligned}$$

Here $\mathcal{A}_{\mathbb{R}^{2n}}$ is the map previously introduced on the loop space.
The critical points of $S(\lambda, z)$ are those pairs (λ, z) such that

$$\begin{aligned} &\int_{S^1} H(z(t)) \cdot dt = c \\ &\dot{z} = \lambda \cdot X_H(z) \end{aligned} \tag{78}$$

Moser uses a different sign convention in defining a Hamiltonian vector field. We see that z is a constant if $\lambda = 0$ and $H(z(0)) = c$. If $\lambda \neq 0$ we can define $w(t) = z(\frac{t}{\lambda})$ and note that $w(0) = z(0) = z(1) = w(\lambda)$. Moreover, the equations for z are equivalent to

$$\frac{dw}{dt} = X_H(w) \quad \text{and} \quad \frac{1}{\lambda} \cdot \int_0^{\lambda} H(w(t)) = c.$$

Using the first equation we find

$$\frac{d}{dt}(H(w(t))) = dH(w(t)) \left(\frac{dw}{dt} \right)(t)) = \omega(X_H(w(t)), X_H(w(t))) = 0$$

since ω_0 is skew-symmetric, i.e. $\omega(A, B) = -\omega(B, A)$. Here ω is the standard symplectic form on $\mathbb{R}^{2n}$. Since $t \to H(w(t))$, as just seen, is constant and the mean value of this function is c we finally obtain

$$\frac{dw}{dt} = X_H(w), \quad w(0) = w(\lambda), \text{ and } \; H(w(t)) = c \text{ for all } t \in \mathbb{R} \tag{79}$$

which just says that if (λ, z) is a critical point of S with $\lambda \neq 0$, then $w : \mathbb{R} \to \mathbb{R}^{2n}$ defined by $w(t) = z\left(\frac{t}{\lambda}\right)$ is a λ-periodic solution of the Hamiltonian system $\dot{w} = X_H(w)$ on the energy surface $H \equiv c$.

44. Moser talks explicitly about a Lyapunov-Schmidt reduction,[**Ly, Schm**], which is a construction which reduces under suitable assumptions an infinite-dimensional equation in a Banach space to a finite-dimensional one. Moser applies this to a local problem and shows that making proper choices the finite-dimensional problem is again variational. Later Amann, [**Am**], and Amann-Zehnder, [**AZ, AZ1**], study this idea in a more global context and a global version is used in the proof of the famous Conley-Zehnder theorem, [**CZ**].

45. The question is: Are there smooth(!) compact regular energy surfaces in $\mathbb{R}^4$ without periodic orbits? This is a longstanding open problem. Joel Fish and Helmut Hofer have shown that a compact smooth energy surface in $\mathbb{R}^4$ has a proper closed invariant subset—which could be a periodic orbit, [**FiH2**].

46. As Rabinowitz writes in [**Ra2**] in the introduction (page 159):
"Much of the motivation for the techniques we use, especially those of Section 2 was provided by our recent paper [11] on free and forced vibrations for semilinear wave equations. Indeed the results of Section 2 can be considered to be the Hamiltonian analogues of [11].
I thank Jürgen Moser who encouraged me to work on the problem of Section 1, Edward Fadell for many discussions on topological matters, and Charles Conley and Michael Crandall for some suggestions." ([11] refers to [**Ra1**].)

47. He carefully picks finite-dimensional linear vector subspaces E_k of E such that

$$E_k \subset E_{k+1} \cdots \subset E$$

and he studies the restrictions of Φ to the E_k, denoted by Φ_k. The first big step is of an analytical nature. He shows the following property, which analysts would call a compactness property.

(CP). If one can find critical points $z_k \in E_k$ of Φ_k so that there exist numbers $0 < a < b < \infty$ with $\Phi_k(z_k) \in [a, b]$ for all large k, then Φ has a critical point $z \in E$ with $\Phi(z) \in [a, b]$.

The compactness property reduces the problem to finding numbers $0 < a < b < \infty$ so that for sufficiently large k one can find critical points z_k of the finite dimensional approximations—that is, to find $z_k \in E_k$ such that

$$d\Phi_k(z_k) = 0 \quad \text{and} \quad \Phi(z_k) \in [a, b].$$

It is completely unavoidable that the Morse-indices $m_{\Phi_k}(z_k)$ have to go to infinity in a quite controlled way. In some sense, for cleverly chosen E_k, Rabinowitz picks a procedure to find critical points $z_k \in E_k$ of Φ_k, where roughly speaking:

$$m_{\Phi_k}(0) \approx \frac{1}{2} \cdot \dim(E_k) \quad \text{and} \quad m_{\Phi_k}(z_k) \approx 1 + \frac{1}{2} \cdot \dim(E_k).$$

This requires nontrivial work and uses the fact that the E_k can be picked in such a way that they are invariant under the action of S^1 on E, coming from the translational action of S^1 on the domain S^1 of the loops. This allows, since Φ is invariant under this S^1-action, to employ the so-called Fadell-Rabinowitz index theory from [**FR**] to show that x_k with the desired properties exist. It can be seen as a generalization of Lusternik-Schnirelman's category theory to invariant subsets for compact Lie group actions.

48. George David Birkhoff is one of the important figures in American mathematics, considered as a representative of American mathematics "coming of age," see [**Mo4**] for a discussion of Birkhoff and his work.

49. Finding geodesics or closed geodesics is a classical problem which already was important in the development of Morse theory. Finding geodesics for Finsler metrics is a generalization of the Riemannian geodesic problem which we explain now.
Consider a smooth compact regular surface Σ in the Euclidean space $(\mathbb{R}^N, d_{\mathrm{Eucl}})$. The inner product on $\mathbb{R}^N$ is denoted by $\langle \cdot, \cdot \rangle$. The length of a smooth curve $\gamma : [a, b] \to \mathbb{R}^N$ is defined by

$$\mathrm{length}(\gamma) = \int_a^b \sqrt{\langle \dot{\gamma}(t), \dot{\gamma}(t) \rangle} \cdot dt.$$

Given two different points in $\mathbb{R}^N$, say x and y, there is precisely one shortest path starting at time $a = 0$ at x, with speed (velocity) 1 and arriving later at y. Namely we move on a straight line from x to y, and since our supposed velocity is 1 we will arrive at time $b = \mathrm{dist}(x, y)$:

$$\gamma : [0, \mathrm{dist}(x, y)] \to \mathbb{R}^N : \gamma(t) = x + \frac{t}{\mathrm{dist}(x, y)} \cdot (y - x).$$

Next assume that the two points lie on Σ and we want to travel from x to y on Σ(!) with speed 1 and we would like to find the (or a) shortest path. For example let

$$\Sigma = \mathbb{S}^1 = \{(x, y) \in \mathbb{R}^2 \mid x^2 + y^2 = 1\}$$

be the unit circle in $\mathbb{R}^2$. Consider the points $(-1, 0)$ and $(1, 0)$ on Σ. The shortest way of travel in $\mathbb{R}^2$ is along the x-axis and the distance is 2. If we require to travel on Σ the distance is π which is half the circumference of the unit circle. More generally if our surface is the unit sphere $\mathbb{S}^{N-1} \subset \mathbb{R}^N$, then given the north pole x and the south pole y there are many shortest paths between x and y and these trace out part of the great circles. It turns out that if x is the north pole and y is neither the north pole and the south pole, there exists precisely one shortest path which traces out part of a great circle containing y.
Instead of measuring the length using $\int_a^b \sqrt{\langle \dot{\gamma}(t), \dot{\gamma}(t) \rangle} \cdot dt$ we can replace $\sqrt{\langle \dot{\gamma}(t), \dot{\gamma}(t) \rangle}$ by other expressions to be integrated. Among the possibilities are Finsler metrics, which we shall not discuss here further.

50. **Weinstein Conjecture:** Let M be a smooth closed manifold of odd dimension $2n+1$ and assume λ is a differential one-form on M such that $\lambda \wedge d\lambda^n$ is a volume form. Then the so-called Reeb vector field X defined by

$$i_X\lambda \equiv 1 \quad \text{and} \quad i_X d\lambda \equiv 0$$

admits a nontrivial periodic orbit.
As a corollary one obtains the following. If W is a symplectic space and $H : W \to \mathbb{R}$ a Hamiltonian we can study

$$\dot{x} = X_H(x), \quad H(x(t)) = E, \quad x(0 = x(T), \ T > 0$$

i.e. we would like to find a T-periodic orbit on the energy surface $\Sigma := \{w \in W \mid H(w) = E\}$. We assume that the energy surface is regular and closed. As a corollary of the Weinstein conjecture we obtain the following criterion. If there exists a one-form λ on Σ satisfying $d\lambda = i^*\omega$, where ω is the symplectic form on W and $i : \Sigma \to W$ the inclusion, then the property that $\lambda \wedge (i^*\omega)^n$ is a volume form on Σ would imply the existence of a periodic orbit for the Hamiltonian system, provided the Weinstein conjecture is true. Here $2n+2 = \dim(W)$.

51. Besides for any contact form on S^3, Hofer's results also covered closed three-manifolds with non-vanishing second homotopy group and arbitrary closed three-manifolds provided the contact structure associated to the Reeb vector field is overtwisted.

52. Hofer, Wysocki and Zehnder introduced in [**HWZ1**] the notion of "dynamical convexity" for an energy surface. This notion is symplectically invariant and captures some of the properties of convexity. In very interesting work by Chaidez and Edtmair, [**ChE**], on this topic, it is proved that convexity and symplectic convexity are inequivalent notions in all dimensions (≥ 3).

53. Ekeland was also a consultant to Jeff Goldblum who played a mathematician specializing in chaos theory in the movie 'Jurassic Park.'

54. An alternative proof was given in [**H5**] by using directly a version of the action functional following ideas from [**H4**] which in turn were extensions of ideas in [**BeRa**].

55. Since the original work in [**HWZ1**, **HWZ2**] this area has flourished with the work by Colin, Contreras, Florio, Hryniewicz, Mazzucchelli, Salomão and Rechtman, to name a few.

56. See the work by Cristofaro-Gardiner and Hutchings, [**CH**], as well as Cristofaro-Gardiner, Hutchings and Pomerleano, [**CHP**], for more results on the two or infinitely many periodic orbits dichotomy. More recent developments based on Hutchings' embedded contact homology and the volume formula, [**CHR**], of Cristofaro-Gardiner, Hutchings and Ramos, led Kei Irie to prove a C^∞-closing lemma for Reeb flows on closed three-dimensional contact manifolds. A consequence of Irie's result is that a generic strictly convex energy surface in $\mathbb{R}^4$ carries a dense set of periodic orbits, and shown even more recently an equidistributed set of periodic orbits, see [**Hut1**, **Hut2**, **Hut3**, **CHR**, **Ir1**, **Ir2**]. One of the current activities investigates the higher dimensional case. A first breakthrough was achieved by Julian Chaidez, Ipsita Datta, Rohal Prasad and Shira Tanny, [**CDPT**].

57. The Conley-Zehnder index is concerned with the homotopical properties of arcs of linear symplectic maps. Namely denote by Sp(2n) the group of linear symplectic maps $\Phi : \mathbb{R}^{2n} \to \mathbb{R}^{2n}$. We denote by $\mathrm{Sp(2n)}^*$ the subset of all those maps which do not have 1 in their spectrum and consider the metrizable space Σ^{2n} of continuous maps

$$\bar{\Phi} : [0,1] \to \mathrm{Sp(2n)} : t \to \bar{\Phi}_t$$

starting at Id and ending in $\mathrm{Sp(2n)}^*$. Then one can consider the space of homotopy classes of paths $H\Sigma_{2n}$ and the Conley-Zehnder index defines a bijection $\mathrm{CZ}_{2n} : H\Sigma_{2n} \to \mathbb{Z}$. The Conley-Zehnder indices CZ_{2n} for the different dimensions $2n$ are interrelated by certain operations and they can be uniquely characterized by a set of axioms. Besides the original paper one should also consult [**RS**].

58. A German diploma thesis at that time is comparable to a master's thesis.

59. The title of his diploma thesis is: "Klassifikation hochzusammenhängender Poincaré-Räume" (Classification of highly connected Poincaré spaces).

60. Of particular importance for him are [**SaU**, **Uh1**, **Tau**]. The ideas in [**SaU**] will also be important for Gromov in [**G**].

61. The original German text: "Nun ergibt sich jedoch die Möglichkeit, durch ein Beschäftigungsverhältnis an der Universität das Studium in Berkeley bis zur Doktorprüfung weiterzufuehren. Da die University of California auf meinem Arbeitsgebiet zu den besten der Welt zählt, ist dies fuer meine berufliche Zukunft von grosser Bedeutung."

62. We have this information from a draft letter with a certain number of corrections (provided to us by Zehnder). We do not know if he later send a formal letter with this content to Stöcker, but it seems likely.

63. These contributions can be found in [**Pro**]. Paul Rabinowitz described a follow-up of his joint work on indefinite functionals with Vieri Benci, [**BeRa**], which started with his breakthrough results on finding periodic orbits for wave equation problems as well as Hamiltonian systems, [**Ra1**, **Ra2**]. Rabinowitz's work is the starting point for a thread leading to the Conley-Zehnder theorem.
Eduard Zehnder explained his proof with Conley of the Arnold Conjecture, [**CZ**], which resulted in their celebrated theorem—considered the breakthrough result that opened up the field of symplectic topology.
Ivar Ekeland reported on his Morse theory for convex Hamiltonian systems, [**Ek1**]. The convexity assumption allowed him to employ a trick that leads to a critical point problem on an infinite-dimensional space, which, up to some technical points, can be studied using more classical variational methods as developed by Palais and Smale. Without this trick the original problem belongs to the class of problems that seemed inaccessible to critical point theory, before a seminal contribution by Rabinowitz. Although convexity is not a symplectic notion, the richness of the theory is surprising.
Alan Weinstein discussed how the Conley-Zehnder result may be extended to more general manifolds and how the 'isolating block' theory (use of the Conley index) can be replaced by more conventional critical point theories of Liusternik-Shnirelman and Morse.

64. Bahri was presumably the first spending a lot of time studying the Weinstein conjecture. His ideas were interesting, [**Ba1**], but did not lead to any significant results concerning the Weinstein conjecture. However, these ideas turned out to be useful in studying elliptic systems. After Hofer's paper on the Weinstein conjecture, [**H0**], he returned to the Weinstein conjecture in [**Ba2**]. Maybe it is worthwhile to reconsider his ideas with the better methods we have today. Bahri had read Daniel Bennequin's Ph.D.-thesis in great detail and Bennequin had thanked him in an acknowledgement. Bennequin's thesis which is important for the development of contact geometry influenced Eliashberg and was later published in the Astérisque, [**Ben**], then translated into Russian, appearing in Uspekhi, and then translated from Russian into English which resulted in Abbas Bahri's name in the acknowledgment suffering some collateral damage—it became Abbé Bari, [**Ben'**].

65. To elaborate, the other important developments start with Thom's 1949 dream, [**Th**], to use the "descending cell decomposition" given by the (negative) gradient flow; followed by Smale's clarification of these dreams by proving the needed transversality results, his handlebody theory, the proof of the generalized Poincaré conjecture, and the h-cobordism theorem. The final contribution in this development is Floer theory.

66. Thom, Smale, and Witten are all Fields Medal recipients. It seems certain that Floer was discussed for a 1990 Fields Medal, but due to his young age he would still qualify in 1994. This was not the case for the 1990 Fields medalists V. Drinfeld, V. Jones, Sh. Mori, and E. Witten, where only Drinfeld could have been perhaps considered as borderline qualifying in 1994. Since the impact of Floer's work was so overwhelming, a Fields Medal for him should have been a certainty at the International Congress of Mathematicians in Zuerich in 1994 (ICM 1994), had he been alive; at least modulo politics, which very often underlies such decisions.

67. Floer must have seen Witten's paper, [**Wi**], before Summer 86, possibly before Summer 1985. He mentions it in the abstract of his talk at the Arbeitstagung in Bonn on June 15th 1986, but not in his research announcement [**F00**], submitted August 1st, 1986, where he just notes that his construction in finite dimensions has been used by Milnor in [**Mi2**]. In the paper [**F2**], page 518, Floer says referring to [**Wi**]: "... which partly prompted our approach to the Morse theory of the symplectic action." According to Viterbo, Floer had told him—while both where at the Courant Institute—that he had been inspired by the Witten paper.

68. See the book by Matthias Schwarz, [**Sch**], for a presentation of Morse homology which uses gradient flow lines to define a homology theory and prove the usual axioms of a homology theory. A further step is taken by Jean-François Barraud, Mihai Damian, Vincent Humilière and Alexandru Oancea in their upcoming book [**BDHO**].

69. The emerging Floer homotopy theory is concerned with these types of questions.

70. In 1983 several papers which are important for the development of symplectic geometry are discussed in the Bourbaki seminar: The Witten paper, Berestycki talked about "Periodic solutions of Hamiltonian systems," [**Bere**], Douady on "Knots and contact structures" (Bennequin's thesis [**Ben**]), [**Dou**], and Chaperon about the Conley-Zehnder proof of the Arnold conjecture, [**Ch2**].

71. Also the asymptotic behavior of f matters in case Λ is not compact.

72. In some sense Conley's theory is a generalization of finite dimensional Morse theory. Rybakowski, [**Ry**], undertook a first attempt to generalize Conley's theory in such a way that it could be viewed as a counterpart to the PS-extension of Morse theory. Other interesting results in this direction are the generalization of the Conley index to discrete dynamical systems by Robbin and Salamon, [**RS2**]. It is presumably a good idea to have a look from today's perspective at these attempts.

73. See the epigraph in [**GM**].

74. The use of the word "natural" here means that the choices that need to be made in a construction can be given by a general recipe leaving no ambiguity. In many mathematical constructions choices have to be made even if the end result is not ambiguous. In this case the overall construction is natural and uses a natural set of choices (albeit not each individual choice is natural).

75. There are other procedures as well, where one picks as "mathDATA" all Morse-Smale pairs (f, g) for the same Λ and then constructs $C(f, g)$ and given an additional (f', g) another $C(f', g)$ and a chain map $C(f, g) \to C(f', g)$ etc.

76. The procedure to define $C(f)$ can essentially be used to show that the end result $H(f)$ does not depend on f, but only on the underlying space Λ. All this requires serious work, see for example [**Sch**].

77. It is unique up to natural isomorphism.

78. Assuming also some compactness.

79. It means mathematicians can prove this.

80. If the manifold is not compact we have to describe a class of Morse functions with a suitable behavior at infinity.

81. A striking example of this is that the non-degeneracy of the so-called Hofer metric, [**H8**] is equivalent to the validity of a generalization of Gromov's non-squeezing result, [**G**]. A fact which seemingly escaped both Gromov and Hofer. The equivalence was established by Lalonde and McDuff in [**LMc1**].

82. As Marston Morse writes in [**Mo4**]: "Poincaré was Birkhoff's true teacher. There is probably no mathematician alive who has explored the works of Poincaré in full unless it be Hadamard, but in the domains of analysis Birkhoff wholeheartedly took over the techniques and problems of Poincaré and carried on." If one is interested in dynamical systems and its history one should read Birkhoff's influential book [**Bi1**]. Of interest is also Birkhoff's presidential address reflecting about American mathematics [**Bi**].

83. A beautiful connection is the old "rectangular peg problem", which was recently solved by Joshua Greene and Andrew Lobb, by connecting it to symplectic geometry, see [**GL**], after important work by Cole Hugelmeyer, [**Hu**].

84. It is not clear when this precisely happened. Gromov was at the State University of New York at Stony Brook a professor from 1974–1981, a professor at the Université de Paris VI, 1981-1982, and a professor at the Institut des Hautes Études Scientifiques since 1982–. Gromov did not remember when he gave the manuscript to the Orsay topology group, Eliashberg thought it happened while Gromov was at IHES, and Laudenbach, a member of the Orsay topology group, could not remember.

85. According to Laudenbach.

86. The relationship between Herman, Katok and Mather was interesting. Katok and Herman admired the mathematical depth of Mather. According to Zehnder, Herman and Katok appreciated each other's work. The appreciation, however, might escape the casual bystander. This story, known by those who attended the biennial Oberwolfach workshop on dynamical systems (organized by Moser and Zehnder) was contributed by Hofer to the memorial tribute "Anatole Katok—A Half-Century of Dynamics" written by Boris Hasselblatt.

Seating plan for the "Diagonal Argument" as explained by Helmut Hofer.

The following is taken from [**Ha**]: "Katok's usual seat was in the front right (sitting in front was his general custom, and the doors being on the right in the Oberwolfach meeting room makes this the natural side for late-comers). Michael Herman, another prominent regular, might sit in the very left rear corner. It was not uncommon for one of Katok's amendments to someone else's talk to be countered by Herman from the opposite corner: 'I can't believe that I am hearing this! I have three counterexamples to that statement. No, I have infinitely many counterexamples to what you are saying...,' and so the argument across the diagonal of the room was on. This is memorable, of course, not only because it is amusing but because these interactions were illuminating for all those present and emblematic of the creative spark of those meetings."

87. One can construct a Morse function on T^{2n} with precisely 2^{2n} critical points and a smooth function with $2n+1$ many points. Then multiplying these functions by a small ε one can show that the associated autonomous Hamiltonian vector field only has the critical points as 1-periodic orbits and they are non-degenerate in the first case. This shows that the estimates are sharp in this case.
To be more precise we first we observe that if X is a smooth vector field on a closed manifold M and $\|X\|_{C^2}$ is small enough, the only 1-period orbits are the rest points. Fix a Riemannian metric with associated distance d and cover M by finitely many charts. Then there exists $\varepsilon > 0$ such that for every $m \in M$ there exists one of these charts so that $\bar{B}_\varepsilon(m)$ is contained in this chart. We require that the C^0-norm of X is less than ε. That is a measurement in (M, g). Pushing forward X to the charts our measurement of the size of the push forwards is that C^0-norm is less than 2π. We need the inequality for a 1-periodic loop in $\mathbb{R}^n$ of mean-value zero.

$$\|x\|_{L^2} \leq 2\pi \cdot \|\dot{x}\|_{L^2}$$

If X is a vector field on $\mathbb{R}^n$ with a universal bound C on its derivative and $\dot{x} = X(x)$, then $\ddot{x} = X'(x)\dot{x}$ and

$$\|\dot{x}\|_{L^2} \leq 2\pi \cdot \|\ddot{x}\|_{L^2} \leq C \cdot \|\dot{x}\|_{L^2}$$

Hence $\dot{x} = 0$ provided $C < 2\pi$.
On T^m we have a Morse function with exactly 2^m critical points, namely

$$f(x_1, \dots, x_m) = \sum_{i=1}^{m} \sin(2\pi x_i)$$

A smooth function on T^2 with 3 critical points arises as follows. Define on $\mathbb{R}^2$ the function g by

$$g(x_1, x_2) = \left(\Pi_{i=1}^2(\sin(\pi \cdot x_i))\right) \sin(\pi \cdot \sum_{i=1}^{2} x_i)$$

This function is smooth and invariant under $\mathbb{Z}^2$ and defines a smooth function on T^2. Further we can permute the coordinates. If x is a critical point we have for every $\ell = 1, 2$

$$(\Pi_{i=1}^2 \sin(\pi \cdot x_i)) \cdot \cos(\pi \cdot \sum_{i=1}^{2} x_i) + \cos(\pi \cdot x_\ell) \cdot \Pi_{i \neq \ell} \sin(\pi \cdot x_i) \cdot \sin(\pi \cdot \sum_{i=1}^{2} x_i) = 0$$

Assume that $g(x) \neq 0$. In this case $x_i \in (0,1)$ and $\Pi_{i=1}^2(\sin(\pi \cdot x_i)) \neq 0$ as well as $\sin(\pi(x_1 + x_2)) \neq 0$ and we infer for $\ell = 1$

$$\cot(\pi \cdot \sum_{i=1}^{2} x_i) + \cot(\pi \cdot x_\ell) = 0$$

and consequently $0 < x_1 = x_2 < 1$. We can write

$$\cot(2\pi x_1) + \cot(\pi \cdot x_1) = 0$$

If $2x_1 < 1$, we find by monotonicity of the above expression precisely one solution with $0 < 2x_1 < 1$. If we assume that $0 < x_1 < 1$ and $1 < 2x_1 < 2$ again there is precisely one solution again.
If now $g(x) = 0$ assume first that $\sin(\pi \cdot \sum_{i=1}^2 x_i) = 0$ implying that $\cos(\pi \cdot \sum_{i=1}^2 x_i) = \pm 1$. In this case $\Pi_{i=1}^2 \sin(\pi \cdot x_i) = 0$. This is only possible if at least one of the factors is zero. From $\sin(\pi \cdot \sum_{i=1}^2 x_i) = 0$ we infer that the other coordinate vanishes as well. Hence we have that $x = (0, 0)$, which is indeed a critical point on the zero level, i.e. $g(0, 0) = 0$.
Assume next that $\sin(\pi \cdot \sum_{i=1}^2 x_i) \neq 0$ implying that $\Pi_{i=1}^2 \sin(\pi \cdot x_i) = 0$ so that at least one of x_1 and x_2 equals 0. For example assume that $x_1 = 0$. This implies that $x_2 = 0$ and contradicts $\sin(\pi \cdot \sum_{i=1}^2 x_i) \neq 0$.
In summary, we find three solutions namely $x = (0, 0)$, $y = (x_1, x_1)$ with $0 < x_1 < 1/2$ and $z = (x_2, x_2)$ with $0 < x_2 < 1 < 2x_2 < 2$. Consequently g induces a function with exactly three critical points on T^2.

88. The paper, which was submitted on March 15th 1983 to inventiones mathematicae, appeared in the February 1983 issue; a preprint started circulating at the beginning of 1983. Refereeing seemed to be much faster in those days.

89. Zehnder is not absolutely sure on which occasion this happened. He believes that he gave a talk on their result in Paris during the time that Conley was in Bochum—from fall 1982 to summer 1983—since he remembers Conley was with him. This places the event in the first half of 1983. Zehnder also mentions that Conley did not like to give talks. (Conley, tragically, committed suicide on November 20th 1984.)

90. Via several interviews with Rolf Kaiser and Rebecca Steinberg.

91. `https://allenginsberg.org/`

92. Via interviews with Kaiser and Rob Kusner.

93. This collaboration resulted in a highly cited paper [**FW**] about the (non-)spreading of wave packets—the work used techniques by Taubes, [**Tau, Tau2**] and Jaffe/Taubes, [**JT**], in Yang-Mills theory (which had been touched on during the Donaldson Seminar) as well as previous work by Weinstein, [**We7**].

94. As Zehnder describes in [**Ze0**]. It is not possible to date this event more precisely than a general time interval. The first talk about the Conley-Zehnder theorem was given by Zehnder at a colloquium at the University of Zürich shortly before Christmas 1982; and Zehnder already knew Floer when he came to the Browder conference in July 1983. Since the Spring semester in Berkeley ended in May, Floer might have briefly visited Bochum around this time. Also it is likely that Floer had already heard about the Conley-Zehnder theorem through Weinstein who would have received a preprint. The Conley-Zehnder paper was submitted to Inventiones Mathematicae on March 15, 1983.

95. As Zehnder describes in [**Ze0**].

96. Hofer gave a talk about the (his) Arnold conjecture in Paris in January 1985 and the paper, [**H1**], had been submitted in November 1984; at the same time preprints had been sent out. In January 1985, Laudenbach and Sikorav [**LS**] submitted a proof of this Arnold conjecture using generating function type methods motivated by contributions by Marc Chaperon, [**Ch1, Ch2**]. The paper [**LS**] is considerably shorter than [**H1**], i.e. 9 pages versus 59 pages. Floer had previously made fun of Hofer's hard work in [**H1**] compared to the easy ride of [**LS**], and that might have been the occasion to admit that he had panicked. There was never such an admission.

97. Gromov highlighted this in his 1986 ICM talk, [**G3**].

98. In an interview, [**G2**], that is part of a Simons Foundation collection, Gromov remembered how he got interested in symplectic geometry and how his views developed. He described the troubles Eliashberg had and his success in finding rigidity. Particularly Gromov mentioned the Conley-Zehnder theorem and the importance of Rabinowitz's idea, which he had not been aware of.

99. Fundamental initial work in this direction (predating Gromov's epochal [**G**]) is Bennequin's Ph.D.-thesis, [**Ben, Ben'**], which later triggered Eliashberg's "overtwisted paper",[**Eli**]. Of course, from our today's perspective, Rabinowitz' breakthrough result, [**Ra2**], can be considered as a hint of the existence of the 'hard side' of symplectic geometry.

100. See also the discussion at the end of Subsection 4.3.

101. Unfortunately, Zehnder destroyed all his letters of reference after retirement. Taubes did not remember if he had written a recommendation letter.

102. At that time Courant was seen by the university's administration as the "jewel in the crown of NYU", while faculty outside of the institute rather viewed the Courant faculty as a group of pampered spoiled brats. As one gauge of the faculty's quality, the Abel Prize (comparable to the Nobel Prize) was given for the first time in 2002, and its recipients over the years have included a number of Courant mathematicians: Peter Lax in 2005; Srivasa Varadhan in 2007; Mikhail Gromov in 2009; Louis Nirenberg in 2015. (Gromov was half-time at Courant otherwise affiliated with the IHES in Bures-sur-Yvette, and Nirenberg shared the prize with John Nash, who had previously received the Nobel prize in economics).

103. As described by Floer, Mark Goresky, Robert MacPherson and Kusner.

104. The first talk for which there is a recording was delivered by Floer on October 29th, 1986 in Stony Brook, [**F10**].

105. Chern-Simons theory is named after Shiing-Shen Chern and James Simons and is a three-dimensional topological quantum field theory.

106. Some of the material in the book [**DK**] should have been floating around several years beforehand and Atiyah would have been aware of it.

107. The conference volume contains articles which do not match precisely the talks given at the conference. For example, Floer was not able to participate due to a visa issue related to his move from SUNY to Courant, but he contributed an article. Zehnder's article describes the almost existence result obtained at the conference and Rabinowitz describes an improvement of this result.

108. At least generically.

109. As Viterbo remembers it: "Andreas, rather upset, wrapped up his talk. There was a few minutes break between the two talks and I forgot what I said to him, but I remember him replying in a laugh 'Yes, you are right, in Germany you cannot even go to the bathroom without having a clock on the wall'."

110. Another by-product of their work at that time is the well-known Conley-Zehnder index. The Conley-Zehnder index turns out to be a substitute for the Morse index in the following sense: For the (bad) action functionals the classical Morse index or Morse co-index is always ∞, but it turns out that there is enough structure to talk about the difference of two such infinite Morse indices. The difference of the Conley-Zehnder indices is then precisely the difference of the two infinities.

111. This is one of the examples how an outside idea can make a difference. Given the state of symplectic geometry before Conley-Zehnder, no global tools were known in the field. Generating functions were known, but not in the sophisticated way as introduced by Marc Chaperon [**Ch1**] following the Conley-Zehnder breakthrough—the action functional is the infinite-dimensional version of a generating function and under certain assumptions the infinite-dimensional functional can be reduced to a finite-dimensional one, for example the saddle point reduction or the Chaperon reduction are ways to carry this out. It is also remarkable that the full strength of Conley and Zehnder's work (before the Arnold conjecture) is not even needed. They only needed to learn about the problem and produced a seminal result which jump started a whole development.

112. While the initial ideas of Palais and Smale aimed at generalizing finite-dimensional Morse theory, Ambrosetti and Rabinowitz were focused on developing these methods having applications to partial differential equations in mind.

113. The book contains a vast amount of material, but not in the form of what one needs. However, it would be possible, although painful, to develop, starting with this book, useful ideas for the problem at hand.

114. The following observation shows that maps u would, assuming they exist, have some nice properties: If we have a bounded flow line in the sense as just defined we obtain from (34) that there exist sequences $s_k \to \pm\infty$ such that $\Phi(u(s_k)) \to d \in \mathbb{R}$ and $\nabla_{L^2}\Phi(u(s_k)) \to 0$ in L^2. If we view $u(s_k)$ as lying in $T^{2n} \times ((P^- + P^+)H^1(S^1, \mathbb{C}^n))$ it follows with not too much effort that after taking a subsequence we may assume that $u(s_k)$ converges to some $x : S^1 \to T^{2n}$ in the Hilbert manifold $H^1(S^1, T^{2n})$ so that x is a contractible periodic orbit. This property can be viewed as a Palais-Smale type condition for bounded flow lines for the unregularized gradient flow. In other words: bounded flow lines for the unregularized gradient flows have something to do with critical points. At this point of the development maybe a reason to investigate the existence of such bounded solutions further.

115. Namely one solves (27) for a sequence of Ks tending to infinity. For each approximation one knows that the "flow lines" converge to critical points which belong to a compact subset. This will imply that we can obtain useful estimates for obtaining "flow line" for the unregularized gradient flow as used by Floer.

116. Witten's paper references Milnor's book on the h-cobordism theorem, [**Mi2**], at one point saying (where [14]=[**Mi2**])
"The effect of tunneling can be calculated in the WKB approximation, or, in a current language, by means of instantons [14]. Tunneling effects often remove spurious degeneracies which exist in perturbation theory, and so it is in this case."
Here "instantons" refers to rigid flow lines in [**Mi2**]. Floer references [**Mi2**] in his research announcement [**F0**], submitted on August 1986, but not [**Wi**]. In [**F0**] he says with respect to his construction "For an analogous construction in finite-dimensional Morse theory, see Milnor [6]", ([6]=[**Mi2**]). He gives a talk at the 26th Mathematische Arbeitstagung in Bonn on June 15th, 1986 with the title "Holomorphic curves and fixed points of symplectic maps." This seems to be the first talk he ever gives about this topic. The proceedings of the Arbeitstagung, [**AT**], contain summaries of the talks written by the authors (handwritten or typed). In his notes Floer has three references: [**Mi2**], [**Wi**] and [**FU**]—rather sparse and incomplete since we know that Gromov's paper, [**G**], was crucial for him. It is possible that Floer finished his Bulletin article in Bochum and mailed it (no submissions on the internet yet!) before the Arbeitstagung to the AMS which might have been closed for July so that the receipt is only in August. Floer told Hofer during the Fall semester at Courant how he managed to give a talk at the Arbeitstagung by first talking to Atiyah who became very interested. From this discussion with Atiyah resulted a new project—instanton Floer homology for 3-manifolds. It is possible that Atiyah pointed out the relevance of [**Wi**]. In the basic technical paper [**F1**] only [**Mi2**] appears, but not [**Wi**]. In [**F2**] Floer writes "It is proved in [9](=[**F3**]) that its cohomology can be computed by a construction analogous to the definition of I^*. This observation, which represents an alternative proof of the Morse inequalities, has been partly used in [16](=[**Mo2**]). With the field $\mathbb{Z}_2$ replaced by $\mathbb{R}$, it has been studied by Witten in his recent paper [27](=[**Wi**]), which partly prompted our approach to the Morse theory of the symplectic action."

117. How, when, and whom Floer references in his papers indicate something between sloppiness and possible intellectual evolution—Floer could indeed be sloppy. Hofer believed and still believes that Floer had consumed this material [**Wi**] at an early stage, but an alternative scenario is possible. Namely that Floer was familiar with a construction in Milnor's book, [**Mi2**], on the h-cobordism theorem and learned, as already mentioned, about Witten's paper only in 1986 through Atiyah. With Floer's background on the Conley index theory, the idea of a Morse decomposition and the Milnor construction one can come up with the idea for Floer homology after being triggered by Gromov's 1985 seminal paper, [**G**]. In one way or the other, Floer knows at the beginning of 1985 that the combinatorics of rigid gradient lines can be used to compute homology.

118. In the following δ is the differential coming from the count of trajectories:
"In a more general situation, Fransoza [4] defined a "connection matrix" recovering the cohomology of the index of S (see also [9]). In fact, the existence of the homomorphism δ follows directly from Fransoza's work. While it is known that the connection matrix is related to flow trajectories, it is only in the case of a smooth Morse-Smale flow on a finite dimensional manifold that it can actually be obtained from those. Our methods in proving Theorem 1 do not use Fransoza's work but only the elementary technique of Morse decompositions as introduced in [1]."

119. In 1987, Hofer received a Sloan Fellowship—he didn't know he'd been nominated, so it was a surprise. Rutgers University had a program called "competitive fellowship award" which allowed researchers to use prize money or awards one hadn't applied for to buy oneself out from teaching for nominal amounts. The Sloan amount was $25k and the price of a course was around $6k. With an already reduced teaching load it allowed Hofer to spend fall 1987 at IAS, spring 1988 at Courant, and fall 1988 at the MSRI in Berkeley. Hofer is forever grateful to Rutgers, particularly the chairman at the time, Joseph D'Atri (1938-1993), who had nominated him. Floer received a Sloan Fellowship in 1989. Receiving the Sloan—and having that extended collaborative research time—created an immense opportunity which accelerated the "symplectic lift-off." Shortly after the award of the Sloan Fellowship, Hofer received a letter from the American televangelist Robert Schuller from Robert Schuller Ministries "...God loves you and so do I" sharing the joy about the award, and providing an "algorithm" how to transfer the award to the ministries ☺ . Hofer's later Courant colleague Jalal Shatah received a Sloan fellowship, and also Schuller's letter.

120. There are many possible choices for a λ which satisfy $d\lambda = \omega$. But the numbers $\int_0^T \lambda(X_H(z(t))) \cdot dt$ do not depend on the specific choice. We can for example consider the collection of numbers

$$\int_0^T \left(\sum_{j=1}^{2n} (-q_j(t) \cdot \frac{\partial H}{\partial q_j}(q(t), p(t)) \right) dt.$$

121. Today the construction would be called S^1-equivariant symplectic homology. This particular work was never written up, but versions of it lead to the posthumous publications on symplectic homology.

122. Berkeley's hiring success with Floer was used approximately a year later in an attempt to hire Eliashberg, when he was allowed to leave the former Soviet Union. In Fall 1988, Eliashberg participated in the MSRI program on symplectic geometry. In need of a permanent job, he'd applied for several positions among them Berkeley and Stanford. Both offered him full professorships; he accepted the better offer, which was Stanford. Eliashberg recalled that the Berkeley chair tried to convince him to come to Berkeley by boasting that people are so eager to join the Berkeley department that they happily accepted a lesser offer, and Floer was provided as an example.

123. Kaiser recounted these instances of Floer's generosity. And he conveyed the following story (which seemed to take place shortly before Floer moved to Bochum), told to him by Cynthia, one of Andreas' Berkeley friends: "Andreas was always giving away almost all his money to people on the street. One night he went to a store, and he only had a dollar to give to some guy outside the store afterwards. So the guy attacked Andreas for not having more money, like he usually gave. Andreas would not let me give him Band-Aids, make tea, call anyone, did not want to talk—nothing. We sat together for about an hour in silence, then I left to go

back to my apartment. That was the last time I saw him alive." According to Kaiser: ".... That was certainly 100% Andreas. He was shocked but he could not let it out. Not very long after this he was back in Germany with the same bruise on his head and did not want to talk about it much either. He only said that this was one of the everyday risks—in spite of taking the caution not to go out alone in the dark."

124. Devaney's and Flaschka's main area of research were dynamical systems; Guillemin worked on symplectic group actions as well as microlocal analysis, spectral theory, mathematical physics; Weinstein was working in differential geometry, mathematical physics, and symplectic geometry.

125. Floer's talks concern the $\mathbb{R}^{2n}$-setting and essentially construct a version of S^1-equivariant symplectic homology for star-shaped domains. He also outlines a proof of the capacity formula for products.

126. Resulting in four posthumous papers, [**FH**, **FHW**, **CFH**, **CFHW**].

127. Floer-type homology theories and symplectic capacity theory are meanwhile important tools for studying symplectic embeddings. Amazingly not too much is known in dimensions at least six—here even the symplectic embedding theory for ellipsoids is only partially understood. The best results have been obtained for four-dimensional ellipsoid embeddings, [**McD4**], where McDuff obtains an optimal result using Hutching's embedded contact homology, see [**Hut3**] for an overview. On the other hand, also in dimension four, embeddings of polydisks into ellipsoids pose difficult problems as the work by Katherine Christianson and Jo Nelson shows, [**CN**].

128. No joke: His secretary's last name in 1989 was Sawyer.

129. During this event, which took place during the first half of 1990, honorary doctorates are awarded to Hans Grauert, Jürgen Moser and Reinhold Remmert. Among the conference participants were: Eliashberg, Ekeland, Floer, Gromov, Hirzebruch, Nirenberg, Siu, Stein, Viterbo and Zehnder.

130. Huckleberry had been on Floer's thesis committee.

131. Part of the deal with the ministry was to designate one of the existing positions which would become vacant in the future due to retirement and to return it to the ministry at that time. Although in general the retiring professor would not have a say in selecting a candidate for the vacant position, there were other factors involved. For example, a particular field might have been represented with impact at the department for a long time. So it seemed good practice to discuss the plans with the person whose position would be given back to the ministry after retirement. As Huckleberry remembered: "Karl Heinz Spallek originally worked in complex analysis, and I knew his work involving interactions with real geometry. So I hoped that by explaining to him something about connections to symplectic geometry and, in general, the significance of Floer's work, he would be honored to have him as his successor. Since Spallek's position was the next available one (among my friends :-)), I decided to talk to him about it—I doubted that we could push it through without Spallek's support. After some time Spallek got the point and supported it. This really helped with the faculty. The second point was the Rector, Kurt Ibsen. I had known him through various fights, but in the end I think we got along. So I simply made an appointment with him, explained to him Floer's level, explained to him that with Floer and Hofer we would then have one of the top teams worldwide in an area that was extremely important, and finally that he could mark Floer's position as the successor of Spallek. It was surprisingly easy to convince him."

132. This meant that the mathematicians voted for him; then the paperwork went to the senate of the university; then through the Rektor's office to the Ministry of Science and Research. All of which is to say that the paper trail went up the "Dienstweg" (official channels).

133. The same ministry had been unable to keep Faltings in Germany in 1985. The "Faltings disaster" worked quite likely in Floer's favor as it did in HH's case.

134. Kaiser mentioned that Hélène had joined Floer in Kyoto; and she had also joined him on a trip to Moscow.

135. The first talk took place on Thursday, May 2nd, 5pm, 1991, at the Humboldt University; and the second talk on Friday, May 3rd, 4pm, 1991, at the Technical University. Thomas Friedrich remembered in 2011 in an email to Robert Schrader: "I still remember the first

talk by Floer in Room 2097 of the main building of the Humboldt University (HU), but less so the second talk. The room was relatively large, but not completely occupied, but pretty full. I think there were about 150 people in the audience, but definitely more than 100. In particular there were many colleagues from applied mathematics from the HU and the theoreticians from the still existing Weierstrass institute, but also some colleagues from applied math at the Weierstrass institute came. Also some colleagues from the Technical and the Free University came. In any case it was one of the those bigger colloquium talks, which practically had not existed in the last ten years."

136. Schrader mentioned that he learned from Ruedi Seiler at the Technical University that Floer had asked if they had a job for him, and that Floer quite likely raised this question after he gave the talks. While Schrader was visiting IAS in November 2011 he also mentioned that Floer had called a few weeks after the talks asking whether there was a job opportunity at the Free University. Schrader only later learned that he wanted to be closer to his brother who lived in Berlin. Having given the talks on May 2nd and May 3rd, the few weeks Schrader was referring to were less than 12 days. Possibly Floer talked to him shortly before his death.

137. The letter was obtained from the archive of IHÉS with the help of Richard Kruel. We thank Detlef Floer, Misha Gromov and the IHÉS for permissions to publish the content of the letter.

138. Hofer remembers that Floer killed himself in the early morning hours. Huckleberry recalls: "I was in Bochum that morning and found out about it then; so it does make sense it happened in the early morning hours. At some point Storch, Skirde and I met to figure out what we had to do, e.g., in terms of an announcement, informing the administration, etc." Hofer did not participate in this meeting.

139. Two pieces of music Andreas was fond of:
Franz Schubert, Der Wegweiser (The Signpost)
`https://www.youtube.com/watch?v=zfNMCajfP94`
Richard Strauss, Beim Schlafengehen (Going to sleep)
`https://www.youtube.com/watch?v=PXQCUYjoUmI`

140. We thank Dr. Paul Lawley for the permission to use his English translation.
`https://lyricstranslate.com/en/der-wegweiser-signpost.html`

141. Posthumous publications:
   - Andreas Floer and Helmut Hofer, Coherent orientations for periodic orbit problems in symplectic geometry, Mathematische Zeitschrift, Vol. 212 No. 1 (1993), 13–38,
   - Andreas Floer and Helmut Hofer, Symplectic homology. I. Open sets in $\mathbf{C}^n$, Mathematische Zeitschrift, Vol 215 (1994), 37-88,
   - Andreas Floer, Helmut Hofer and Kris Wysocki, Applications of symplectic homology I, Mathematische Zeitschrift Vol. 217 No 4 (1994), 577–606.
   - Kai Cieliebak, Andreas Floer and Helmut Hofer, Symplectic homology II: A general construction, Mathematische Zeitschrift, Vol. 218 No. 1 (1995), 103-122.
   - Kai Cieliebak, Andreas Floer, Helmut Hofer and Kris Wysocki, Applications of symplectic homology. II. Stability of the action spectrum, Mathematische Zeitschrift, Vol. 223 No. 1(1996), 27-45.
   - Andreas Floer, Helmut Hofer and Dietmar Salamon, Transversality in elliptic Morse theory for the symplectic action, Duke Mathematical Journal, Vol. 80 No.1 (1995), 251-292.
   - Andreas Floer, The configuration space of Yang-Mills-Higgs theory of asymptotically flat manifolds. The Floer memorial volume, 43–75, Progr. Math., 133, Birkhäuser, Basel, 1995.
   - Andreas Floer, Monopoles on asymptotically flat manifolds. The Floer memorial volume, 3–41, Progr. Math., 133, Birkhänouser, Basel, 1995.

142. Given a symplectic manifold (M, ω) Gromov showed the importance of the space of smooth almost complex structures J on M compatible with ω. We say that J is compatible with ω provided $g_J := \omega \circ (\mathrm{Id} \oplus J)$ is a Riemannian metric for M. We note that

$$g_J(Jh, Jk) = \omega(Jh, JJk) = \omega(Jh, -k) = \omega(k, Jh) = g_J(k, h) = g_J(h, k),$$

so that $J : T_mM \to T_mM$ is fiber-wise isometric with respect to g_J. The space $\mathcal{J}_\omega$ of smooth compatible almost complex structures is a contractible topological space. Associated to $J \in \mathcal{J}_\omega$ and a Riemann surface (S, j) we have the nonlinear Cauchy-Riemann type operator

$\bar{\partial}_{J,j}$ which associates to a map $u : S \to M$ the map

$$\bar{\partial}_{J,j}(u) = \frac{1}{2} \cdot [Tu + J(u) \circ Tu \circ j] .$$

We note that for $z \in S$ we obtain a complex anti-linear map

$$\bar{\partial}_{J,j}(u)(z) : (T_z S, j) \to (T_{u(z)} M, J)$$

Gromov showed in [**G**] how fruitful the study of this partial differential equation is with respect to understand global symplectic geometry questions.

For Floer's work certain transversality results are needed—one needs to pick J in such a way that $\bar{\partial}_{J,j}(u)(z)$ is generic, i.e. typical in some sense. Of course, one first has to define what typical means and then has to show that in some sense 'most' J in $\mathcal{J}$ have these typical properties. These genericity results are in general extremely tricky, see for example the celebrated 'Super-rigidity' result recently proved by Chris Wendl and its history. Floer, having certain properties in mind, wants to show that typical smooth J exist. Collections of smooth functions usually do not define Banach spaces or Banach manifolds—only functions with some finite regularity are needed—and one needs the structure of the Banach spaces to show that typical elements exist.

Floer came up with a new technical device, the "Floer ε-spaces." He associates to a sequence $\varepsilon = (\varepsilon_i)$ of positive numbers a subset $\mathcal{J}^{\varepsilon}_{\omega}$ of the set of smooth compatible almost complex structures $\mathcal{J}_{\omega}$. He is well aware that for many choices of sequences ε the subset $\mathcal{J}^{\varepsilon}_{\omega}$ will be empty. However, in [**F1**] which is his fundamental technical paper he shows that there are carefully chosen sequences ε for which $\mathcal{J}^{\varepsilon}_{\omega}$ is nonempty, reasonably large and contains many elements with the typical properties, see Lemma 5.2 in [**F1**].

143. Seminar participants collectively taught themselves, each tackling and presenting an aspect of the subject, working from the book, "Morse Theory and Floer Homology" by Michèle Audin and Mihai Damian, [**AuDa**].

144. It is also known to be isomorphic to versions of Seiberg-Witten Floer theory and Heegaard Floer theory.

145. Through surveys by McDuff, of Barnard College, [**McD3**], Salamon, [**Sa1**] and others, and readable papers by Hofer and Salamon, [**HS1**].

146. Hofer had been the referee of Floer's technical paper on the unregularized gradient flow, [**F1**], a paper which deals with some of the hard analysis problems which needed to be resolved.

147. Looking back at the development of Floer theory, we see two seminal starting points: the Rabinowitz breakthrough in the world of variational theories and Eliashberg's paper (even if we haven't seen yet the correct version). Out of the geometric "Eliashberg and Gromov world" ultimately pseudoholomorphic curve theory arises and possibly (quite likely) this would have been the case independently if there was the Rabinowitz breakthrough or not. In fact, Gromov only became aware of Rabinowitz' contribution after he had written [**G**], while he was aware of [**CZ**], which relied on ideas which had been developed after Rabinowitz had opened up a new research direction. This research direction had developed new methods, but the applicability was limited to certain types of problems and setups—the reason that Conley and Zehnder could not go beyond the torus case. Gromov's pseudoholomorphic curve theory did not have this limitation, but also did not have the variational spirit of the Rabinowitz direction. There was a big gulf between the two approaches. Not only on a level of viewpoints but also on the technical level. One needed to embrace the variational ideas of the Rabinowitz approach (still to be modified by Witten's ideas) and to build a technical machinery based on drastic modifications of Gromov's ideas—breaking the symmetry. Gromov's breakthrough immediately made it possible to go after some "low-hanging geometric fruits" resulting in a rather immediate further separation between the "Rabinowitz world" and the "Eliashberg and Gromov world." As we know from Gromov's reaction to Floer theory he was unlikely to bridge the divide.

So it was in some sense a lucky incident that via Herman and Mather the awareness about the Arnold conjectures in the "right camp" was created and that Conley and Zehnder had developed methods which could be adapted. Then Floer being adventurous went to Berkeley as a graduate student and immersed himself into what it could offer: learning about gauge theoretic methods and for him, the later technically important "bubbling-off analysis" from

Karen Uhlenbeck and the "gluing analysis" from Taubes. In addition, luckily he had learned that rigid flow-lines for a gradient flows can be used to define homology and then there is the working seminar on Gromov's preprint. Many different threads come together—an opportunity arises—and Floer is ready to take advantage. This being said, nothing is obvious at that time, in particular since enormous technical obstacles have to be surmounted.

148. Non-degeneracy means that for a fixed point $m \in M$ of Ψ the map $\mathrm{Id} - T\Psi(m) : T_mM \to T_mM$ does not have 1 in its spectrum.

149. The properties of the curvatures might be locally different in the Riemannian case.

150. This information is captured by so-called symplectic capacities introduced in [**EH1**, **EH2**]. The theory of symplectic capacities was prompted by Gromov's celebrated non-squeezing result obtained in [**G**]. Meanwhile symplectic capacities are an important tool in symplectic geometry and topology. Many questions are being answered by constructing and computing a suitable capacity. More details about this general area can be found in [**CHLS**].

151. A version already appeared in a Comptes Rendus note in 1965, [**Ar3**].

152. It is a general fact that a homology group with coefficients in a field is automatically a vector space over this field. A homology group with coefficients in a ring R is automatically a module over this Ring. For more details see for example [**Dol1**] or [**Vic**].

153. There is an assumption here: the gradient flow with respect to the metric has to be Morse-Smale, a generic property.

154. Meanwhile, all these problems have been addressed on some level, and there are no topological assumptions one needs to impose. This is true if one uses Floer homology with rational coefficients. The most recent work, which still is being scrutinized, introduces a Floer homology with integer coefficients, which in certain cases gives better estimates.

155. Consider the following model for the two-sphere. We take the vector space $\mathbb{R} \times \mathbb{C}$ with coordinates (a, z) and define

$$\mathbb{S}^2 = \{(a, z) \in \mathbb{R} \times \mathbb{C} \mid a^2 + |z|^2 = 1\}.$$

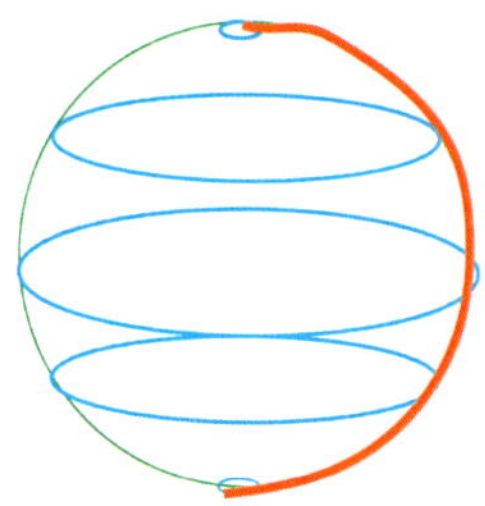

The non-contractible loop γ. For $\tau \in [0, 1/2]$ we move up through the blue loops and for $\tau \in [1/2, 1]$ down through the bold red (constant) loops.

We define a nontrivial continuous loop $\gamma : S^1 \to C^\infty(S^1, \mathbb{C})$ as follows. For $\tau \in [0, 1/2]$ we put

$$\gamma(\tau)(t) = (-1 + 4 \cdot \tau, \sqrt{1 - (-1 + 4 \cdot \tau)^2} \cdot e^{2\pi i t}).$$

For $\tau \in [1/2, 1]$ we define

$$\gamma(\tau)(t) = \left(-2 \cdot (\tau - 1/2) + (1 - 2 \cdot (\tau - 1/2)), \sqrt{1 - (-2 \cdot (\tau - 1/2) + (1 - 2\,(\tau - 1/2)))^2}\right).$$

The map γ starts at time $\tau = 0$ with the constant loop being the South pole. Then the loops get bigger, and at $\tau = 1/4$ we have reached the equator. After this, the loops get smaller, and for $\tau = 1/2$ we have the constant loop at the North pole. From $\tau = 1/2$ to $\tau = 1$, we transport the North pole loop through constant loops to the South pole loop. Then one can show that γ is a non-contractible loop in the space of contractible loops. See the above figure.

156. Naturally, up to homotopy. More precisely, given a symplectic manifold (M, ω), there is a distinguished class of linear maps $J(m) : T_m M \to T_m M$ such that $J(m)^2 = -\mathrm{Id}_{T_m M}$. A smooth selection $m \to J(m)$ of such maps is called an almost complex structure. Given ω there is a subclass of smooth almost complex structures, which are distinguished by the additional requirement that $\omega \circ (\mathrm{Id} \oplus J)$ defines a Riemannian metric. We say that J is compatible with ω. The space $\mathcal{J}$ of all such compatible smooth J is a contractible topological space. These compatible almost complex structures rose to prominence in symplectic geometry through the work of Gromov [**G**].

157. Appropriate choice of Sobolev spaces etc.

158. The book [**GM**] is a readable introduction to these ideas.

159. The exponential decay depends on the size of a spectral gap of an unbounded self-adjoint operator associated to the periodic orbit. This operator is the substitute for the Hessian in the current context. The spectrum also determines up to some normalization the Conley-Zehnder index.

160. In our present context, one should not read too much into it, except that we use some particular conventions in symplectic geometry in the background. It is noteworthy that in symplectic geometry and other fields people often use different conventions. One needs to be aware of this, especially when one relies and combines results of different groups. For example, concerning orientations there are quite often different conventions and one needs to be careful. Such orientation questions also impact the definition of the Conley-Zehnder index.

161. For those using Floer theory it is always important to know the precise conventions. In fact there are many different conventions around covering orientations, definition of Conley-Zehnder indices, the action functional, etc. One has to be aware that combining results originating from different authors can be messy—one possibly needs to translate them into a framework with underlying different conventions.

Acknowledgements

Many thanks to many people:

Central to our understanding of Floer's life are Detlef Floer, Rolf Kaiser and Rebecca Steinberg.

We profited immensely talking to Sir Michael Atiyah, Edward Belbruno, Nathaniel Bottman, Julian Chaidez, Marc Chaperon, Alain Chenciner, Kai Cieliebak, Ivar Ekeland, Yasha Eliashberg, Daniel Freed, Misha Gromov, Alan Huckleberry, Michael Hutchings, Robert Kusner, François Laudenbach, Jacob Lurie, Dusa McDuff, Robert A. Palais, Leonid Polterovich, Rohil Prasad, Peter Sarnak, Robert Schrader, Ruedi Seiler, Shira Tanny, Clifford Taubes, Karen Uhlenbeck, Akshay Venkatesh, Claude Viterbo, Alan Weinstein, Eduard and Jeannette Zehnder, and others.

We also thank Kai Cieliebak, Caden Farley, Brayan Ferreira, Martin Fransson, Diana Gillooly, Michael Kendall, Dusa McDuff, Jo Nelson, Peter Sarnak, Felix Schlenk, Claude Viterbo, Alan Weinstein and Eduard Zehnder for their feedback on early drafts. A special thank-you to our AMS editor Ina Mette, a longtime and steadfast enthusiast; and to the anonymous referees and the AMS team.

We thank the Institute for Advanced Study for generous spacetime support—in particular, the directors Peter Goddard (when the seeds were planted), Robbert Dijkgraaf (during development), and David Nirenberg (through realization).

Many thanks also to the Princeton University Mathematics Department, its chairman Christopher Skinner; the Institut des Hautes Études Scientifiques, its director Emmanuel Ullmo, Claire Lenz, and Richard Kruel for assistance in archival matters.

Further, many thanks to the School of Mathematics at IAS for financial support, and to Nicole Maldonado and Suparna Mahableshwarkar for logistical support.

Many thanks to Caitlin Rizzo for her help on archival and copyright matters. Also many thanks to the Oberwolfach Photo Collection.

We thank Renée Robinson, Yann Blanchet and their crew for "Building Bridges". We enjoyed the deep philosophical discussions with Wayne Leming and Scott Mancini.

Finally, much gratitude and appreciation to the fellow travelers Yasha Eliashberg, Ivar Ekeland, Misha Gromov, Dusa McDuff, Paul Rabinowitz, Dietmar Salamon, Claude Viterbo, Alan Weinstein and Edi Zehnder.

Also, of course, we are indebted to Sonja and Douglas.

Image Credits

xiv Author: Landon Clay, painting, "The Mathematical Jungle" (Cover, Prologue, Fig 1)

xv Source: Chan-Ho Suh; Wikimedia—public domain; Andreas Floer (Prologue, Fig 2)

xviii Source: Detlef Floer; Carte Orange (Prologue, Fig 3)

3 Author: Konrad Jacobs. Source: Archives of the Mathematisches Forschungsinstitut Oberwolfach; Morse—MFO 2930 (Ch 1, Fig 1)

4 Author: Helmut Hofer (Ch 1, Fig 2)

4 Author: Helmut Hofer (Ch 1, Fig 3)

5 Author: Helmut Hofer (Ch 1, Fig 4)

5 Author: Helmut Hofer (Ch 1, Fig 5)

7 Author: Helmut Hofer (Ch 1, Fig 6)

8 Author: Helmut Hofer (Ch 1, Fig 7)

9 Author: Helmut Hofer (Ch 1, Fig 8)

9 Author: Helmut Hofer (Ch 1, Fig 9)

10 Author: Helmut Hofer (Ch 1, Fig 10)

10 Author: Helmut Hofer (Ch 1, Fig 11)

12 Author: Helmut Hofer (Ch 1, Fig 12)

12 Author: Helmut Hofer (Ch 1, Fig 13)

13 Author: Helmut Hofer (Ch 1, Fig 14)

14 Author: Helmut Hofer (Ch 1, Fig 15)

16 Author: Unknown, public domain; Noether (Ch 1, Fig 16)

16 Author: George M. Bergman. Source: Archives of the Mathematisches Forschungsinstitut Oberwolfach; Palais—MFO 5725, cropped (Ch 1, Fig 17)

16 Author: George M. Bergman. Source: Archives of the Mathematisches Forschungsinstitut Oberwolfach; Smale—MFO 14011 (Ch 1, Fig 17)

20 Author: Helmut Hofer; IAS blackboard (Ch 2, Fig 1)

22 Author: Helmut Hofer (Ch 2, Fig 2)

22 Author: Helmut Hofer (Ch 2, Fig 3)

23 Author: Helmut Hofer (Ch 2, Fig 4)

24 Author: Helmut Hofer (Ch 2, Fig 5)

28 Author: Mogens Trolle. Source: `www.mogenstrolle.com`; snow monkey (Ch 2, Fig 6)

31 Author: Gerd Fischer. Source: Archives of the Mathematisches Forschungsinstitut Oberwolfach; Moser—MFO 8842 (Ch 3, Fig 1)

32 Copy from [**Mos**]; fair use; philosphical error (Ch 3, Fig 2)
34 Source: Paul Rabinowitz, private photo; Rabinowitz (Ch 3, Fig 3)
35 Author: George M. Bergman. Source: Archives of the Mathematisches Forschungsinstitut Oberwolfach; Weinstein—MFO 9617 (Ch 3, Fig 4)
36 Source: Ivar Ekeland, private photo; Ekeland (Ch 3, Fig 5)
37 Author: Helmut Hofer (Ch 3, Fig 6)
38 Author: Jürgen Pöschel. Source: Archives of the Mathematisches Forschungsinstitut Oberwolfach; Long—MFO 10018 (Ch 3, Fig 7)
38 Author: Monika Krichel. Source: Erman Cineli; Cineli (Ch 3, Fig 8)
38 Source: Viktor Ginzburg, private photo; Ginzburg (Ch 3, Fig 8)
38 Source: Basak Gürel, private photo; Gürel (Ch 3, Fig 8)
39 Source: Archives of the Mathematisches Forschungsinstitut Oberwolfach; Artstein-Avidan—MFO 20742 (Ch 3, Fig 9)
39 Author: Dan Komoda. Source: IAS; Ostrover (Ch 3, Fig 9)
39 Author: Andrea Kane. Source: IAS; Haim-Kislev (Ch 3, Fig 9)
40 Author: Bob Sine. Source: Charles Conley Jr; Conley (Ch 3, Fig 10)
40 Author: Gerd Fischer. Source: Archives of the Mathematisches Forschungsinstitut Oberwolfach; Zehnder—MFO 8856 (Ch 3, Fig 10)
45 Source: Detlef Floer; baby Andreas (Ch 4, Fig 1)
46 Source: Detlef Floer; Andreas and family (Ch 4, Fig 2)
46 Author: George M. Bergman. Source: `https://commons.wikimedia.org/wiki/File:Shiing-shen_Chern_1977_(borderless).jpg`. Creative Commons Attribution-Share Alike 4.0 International; Chern (Ch 4, Fig 3)
47 Author: Dirk Ferus. Source: Archives of the Mathematisches Oberwolfach; Atiyah and Yau—MFO 8288 (Ch 4, Fig 4)
48 Author: School of Mathematics. Source: Shelby White and Leon Levy Archives Center, Institute for Advanced Study (Princeton , NJ); Donaldson (Ch 4, Fig 5)
48 Author: George M. Bergman. Source: Archives of the Mathematisches Forschungsinstitut Oberwolfach; Uhlenbeck—MFO 13027 (Ch 4, Fig 6)
48 Author: George Bergman. Source: Archives of the Mathematisches Forschungsinstitut Oberwolfach; Freedman—MFO 5121 (Ch 4, Fig 6)
48 Author: School of Mathematics. Source: Shelby White and Leon Levy Archives Center, IAS (Princeton, NJ); Freed (Ch 4, Fig 6)
48 Source: Detlef Floer; Andreas Floer (Ch 4, Fig 7)
50 Author: Randall Hagadorn. Source: Shelby White and Leon Levy Archives Center, Institute for Advanced Study (Princeton, NJ); Witten (Ch 4, Fig 8).
53 Author: Helmut Hofer. Source: Shelby White and Leon Levy Archives Center, Institute for Advanced Study. Photos by Alan Richards, Milnor, Smale; photo by Randall Hagadorn, Witten; photo via Morse family, Thom; (Ch 5, Fig 1)

54 Author: Helmut Hofer (Ch 5, Fig 2)

57 Author: Helmut Hofer (Ch 5, Fig 3)

58 Author: Helmut Hofer (Ch 5, Fig 4)

58 Author: Helmut Hofer (Ch 5, Fig 5)

59 Author: Helmut Hofer (Ch 5, Fig 6)

60 Author: Helmut Hofer (Ch 5, Fig 7)

60 Author: Helmut Hofer (Ch 5, Fig 8)

61 Author: Helmut Hofer (Ch 5, Fig 9)

61 Author: Helmut Hofer (Ch 5, Fig 10)

63 Author: Helmut Hofer (Ch 5, Fig 11)

63 Author: Helmut Hofer (Ch 5, Fig 12)

67 Author: Jürgen Moser; Arnold (Ch 6, Fig 1)

69 Author: George M. Bergman. Source: Creative Commons Attribution-Share Alike 4.0 International, no change; Eliashberg (Ch 6, Fig 2)

69 Author: Dirk Ferus. Source: Archives of the Mathematisches Forschungsinstitut Oberwolfach; Gromov—MFO 8240 (Ch 6, Fig 2)

69 Source: Private photo, Creative Commons Attribution-Share Alike 3.0, no change; Katok (Ch 6, Fig 2)

70 Source: Princeton University, Office of Communications; Mather (Ch 6, Fig 3)

70 Author: Jürgen Poeschel; Herman (Ch 6, Fig 3)

71 Author: Gert-Martin Greuel. Source: Archives of the Mathematisches Forschungsinstitut Oberwolfach; Hingston—MFO 19210 (Ch 6, Fig 4)

79 Source: Kurt Heim and Rolf Kaiser. Creative Commons Attribution-Share Alike 3.0 Unported, no change; Andreas Floer (Ch 7, Fig 1)

79 Author: Andrea Kane. Source: IAS; Murphy (Ch 7, Fig 2)

80 Author: Chelsea L. McCord; McDuff (Ch 7, Fig 3)

81 Author: Christina Buchmann. Source: ETH Zürich; Salamon (Ch 7, Fig 4)

81 Source: François Lalonde, private photograph; Lalonde (Ch 7, Fig 4)

81 Author: Renate Schmid. Source: Archives of the Mathematisches Forschungsinstitut Oberwolfach; Polterovich—MFO 8355 (Ch 7, Fig 4)

83 Author: Gert-Martin Greuel. Source: Archives of the Mathematisches Forschungsinstitut Oberwolfach, Atiyah and Hirzebruch—MFO 12827 (Ch 8, Fig 1)

84 Author: Florence Terrasse; Viterbo (Ch 8, Fig 2)

84 Author: Jeannette Zehnder; Zehnder and Hofer (Ch 8, Fig 3)

85 Author: Konrad Jacobs. Source: Archives of the Mathematisches Forschungsinstitut Oberwolfach; Nirenberg—MFO 3062 (Ch 8, Fig 4)

90 Author: Helmut Hofer (Ch 9, Fig 1)

103 Author: Suny Kim. Source: IAS; Cieliebak (Ch 11, Fig 1)

105 Author: Detlef Floer; Andreas Floer (Ch 11, Fig 2)

107 Author: Sonja Hofer; signpost (Ch 11, Fig 3)

117 Source: Floer Nachlass; Qualifying Exam (Appendix B)
119 Source: Max Planck Institute für Mathematik, Bonn, fair use; Program Arbeitstagung 1986 and Floer talk summary (Appendix C)
123 Source: Floer Nachlass; Progress report (Appendix D)
130 Author: Helmut Hofer (Appendix F, Fig 1)
131 Author: Helmut Hofer (Appendix F, Fig 2)
152 Author: Helmut Hofer (Appendix J, Fig 1)
153 Author: Helmut Hofer (Appendix J, Fig 2)
155 Author: Helmut Hofer (Appendix J, Fig 3)
170 Author: Boris Hasselblatt (Endnote 86, Fig)
178 Author: Helmut Hofer (Endnote 155, Fig)

Bibliography

[AM] Alberto Abbondandolo and Pietro Majer, *When the Morse index is infinite*, Int. Math. Res. Not. **71** (2004), 3839–3854, DOI 10.1155/S1073792804142050. MR2104477

[Ab] D. Abbott, The Reasonable Ineffectiveness of Mathematics [Point of View], Proceedings of the IEEE, Vol 101, No. 10 (2013), 2147–2153.

[AbB] M. Abouzaid and A. Blumberg, Arnold Conjecture and Morava K-theory, `arXiv:2103.01507`.

[AbM] Ralph Abraham and Jerrold E. Marsden, *Foundations of mechanics*, Benjamin/Cummings Publishing Co., Inc., Advanced Book Program, Reading, MA, 1978. Second edition, revised and enlarged; With the assistance of Tudor Raţiu and Richard Cushman. MR515141

[Am] Herbert Amann, *Saddle points and multiple solutions of differential equations*, Math. Z. **169** (1979), no. 2, 127–166, DOI 10.1007/BF01215273. MR550724

[Am1] Herbert Amann, *Ordinary differential equations*, De Gruyter Studies in Mathematics, vol. 13, Walter de Gruyter & Co., Berlin, 1990. An introduction to nonlinear analysis; Translated from the German by Gerhard Metzen, DOI 10.1515/9783110853698. MR1071170

[AZ] Herbert Amann and Eduard Zehnder, *Periodic solutions of asymptotically linear Hamiltonian systems*, Manuscripta Math. **32** (1980), no. 1-2, 149–189, DOI 10.1007/BF01298187. MR592715

[AZ1] H. Amann and E. Zehnder, *Nontrivial solutions for a class of nonresonance problems and applications to nonlinear differential equations*, Ann. Scuola Norm. Sup. Pisa Cl. Sci. (4) **7** (1980), no. 4, 539–603. MR600524

[AR] Antonio Ambrosetti and Paul H. Rabinowitz, *Dual variational methods in critical point theory and applications*, J. Functional Analysis **14** (1973), 349–381, DOI 10.1016/0022-1236(73)90051-7. MR370183

[AT] Arbeitstagung Bonn 1986, `https://archive.mpim-bonn.mpg.de/id/eprint/1081/`.

[Ar1] V. I. Arnol′d, *Mathematical methods of classical mechanics*, Graduate Texts in Mathematics, vol. 60, Springer-Verlag, New York-Heidelberg, 1978. Translated from the Russian by K. Vogtmann and A. Weinstein. MR690288

[Ar2] V. I. Arnol′d, *Matematicheskie metody klassicheskoĭ mekhaniki* (Russian), Izdat. "Nauka", Moscow, 1974. MR474390

[Ar3] Vladimir Arnol′d, *Sur une propriété topologique des applications globalement canoniques de la mécanique classique* (French), C. R. Acad. Sci. Paris **261** (1965), 3719–3722. MR193645

[AKO] Shiri Artstein-Avidan, Roman Karasev, and Yaron Ostrover, *From symplectic measurements to the Mahler conjecture*, Duke Math. J. **163** (2014), no. 11, 2003–2022, DOI 10.1215/00127094-2794999. MR3263026

[AuDa] Michèle Audin and Mihai Damian, *Morse theory and Floer homology*, Universitext, Springer, London; EDP Sciences, Les Ulis, 2014. Translated from the 2010 French original by Reinie Erné, DOI 10.1007/978-1-4471-5496-9. MR3155456

[AFKM] C. Aydin, U. Frauenfelder, D. Koh and A. Moreno, Symplectic methods in space mission design, `arXiv:2308.03391`.

[Ba1] A. Bahri, *Pseudo-orbits of contact forms*, Pitman Research Notes in Mathematics Series, vol. 173, Longman Scientific & Technical, Harlow; copublished in the United States with John Wiley & Sons, Inc., New York, 1988. MR961252

[Ba2] Abbas Bahri, *Flow lines and algebraic invariants in contact form geometry*, Progress in Nonlinear Differential Equations and their Applications, vol. 53, Birkhäuser Boston, Inc., Boston, MA, 2003, DOI 10.1007/978-1-4612-0021-5. MR2002981

[BX] S. Bai and G. Xu, Arnold conjecture over intergers, `arXiv:2209.08599`.

[BDHO] Jean-François Barraud, Mihai Damian, Vincent Humilière, and Alexandru Oancea, *Morse homology with differential graded coefficients*, Progress in Mathematics, vol. 360, Birkhäuser/Springer, Cham, [2025] ©2025, DOI 10.1007/978-3-031-88020-9. MR4917573

[Ben] Daniel Bennequin, *Entrelacements et équations de Pfaff* (French), Third Schnepfenried geometry conference, Vol. 1 (Schnepfenried, 1982), Astérisque, vol. 107, Soc. Math. France, Paris, 1983, pp. 87–161. MR753131

[Ben'] D. Bennequin, Linkings and Pfaff's equations, Russian Mathematical Surveys, Volume 44 (1989), Number 3, 1-65.

[Be1] Edward Belbruno, *Fly me to the moon*, Princeton University Press, Princeton, NJ, 2007. An insider's guide to the new science of space travel; With a foreword by Neil deGrasse Tyson. MR2391999

[Be2] Edward Belbruno, *Capture dynamics and chaotic motions in celestial mechanics*, Princeton University Press, Princeton, NJ, 2004. With applications to the construction of low energy transfers; With a foreword by Jerry Marsden. MR2029316

[BFvK] Edward Belbruno, Urs Frauenfelder, and Otto van Koert, *A family of periodic orbits in the three-dimensional lunar problem*, Celestial Mech. Dynam. Astronom. **131** (2019), no. 2, Paper No. 7, 22, DOI 10.1007/s10569-019-9882-8. MR3905475

[BeRa] Vieri Benci and Paul H. Rabinowitz, *Critical point theorems for indefinite functionals*, Invent. Math. **52** (1979), no. 3, 241–273, DOI 10.1007/BF01389883. MR537061

[Bere] Henri Berestycki, *Solutions périodiques de systèmes hamiltoniens* (French), Bourbaki seminar, Vol. 1982/83, Astérisque, vol. 105, Soc. Math. France, Paris, 1983, pp. 105–128. MR728984

[Ber] Marcel Berger, *Geometry. I*, Universitext, Springer-Verlag, Berlin, 1987. Translated from the French by M. Cole and S. Levy, DOI 10.1007/978-3-540-93815-6. MR882541

[Bey] Nina Byers, *E. Noether's discovery of the deep connection between symmetries and conservation laws*, The heritage of Emmy Noether (Ramat-Gan, 1996), Israel Math. Conf. Proc., vol. 12, Bar-Ilan Univ., Ramat Gan, 1999, pp. 67–81. MR1665436

[Bi] G. D. Birkhoff, Fifty years of American mathematics, Semicentennial Addresses of Amer. Math. Soc.,1938, p. 307.

[Bi1] George D. Birkhoff, *Dynamical systems*, American Mathematical Society Colloquium Publications, Vol. IX, American Mathematical Society, Providence, RI, 1966. With an addendum by Jurgen Moser. MR209095

[Bi2] George D. Birkhoff, *Proof of Poincaré's geometric theorem*, Trans. Amer. Math. Soc. **14** (1913), no. 1, 14–22, DOI 10.2307/1988766. MR1500933

[B3] George D. Birkhoff, *On the periodic motions of dynamical systems*, Acta Math. **50** (1927), no. 1, 359–379, DOI 10.1007/BF02421325. MR1555257

[Bis] Jean-Michel Bismut, *The Witten complex and the degenerate Morse inequalities*, J. Differential Geom. **23** (1986), no. 3, 207–240. MR852155

[Br] L. E. J. Brouwer, *Über Abbildung von Mannigfaltigkeiten* (German), Math. Ann. **71** (1911), no. 1, 97–115, DOI 10.1007/BF01456931. MR1511644

[BZS] Ju. G. Borisovič, V. G. Zvjagin, and Ju. I. Sapronov, *Nonlinear Fredholm mappings, and Leray-Schauder theory* (Russian), Uspehi Mat. Nauk **32** (1977), no. 4(196), 3–54, 287. MR464282

[Bo1] Raoul Bott, *Lectures on Morse theory, old and new*, Bull. Amer. Math. Soc. (N.S.) **7** (1982), no. 2, 331–358, DOI 10.1090/S0273-0979-1982-15038-8. MR663786

[Bo2] Raoul Bott, *Morse theory indomitable*, Inst. Hautes Études Sci. Publ. Math. **68** (1988), 99–114 (1989). MR1001450

[BT] Raoul Bott and Loring W. Tu, *Differential forms in algebraic topology*, Graduate Texts in Mathematics, vol. 82, Springer-Verlag, New York-Berlin, 1982. MR658304

[BouO1] Frédéric Bourgeois and Alexandru Oancea, *The Gysin exact sequence for S^1-equivariant symplectic homology*, J. Topol. Anal. **5** (2013), no. 4, 361–407, DOI 10.1142/S1793525313500210. MR3152208

[BouO2] Frédéric Bourgeois and Alexandru Oancea, *S^1-equivariant symplectic homology and linearized contact homology*, Int. Math. Res. Not. IMRN **13** (2017), 3849–3937, DOI 10.1093/imrn/rnw029. MR3671507

[Ca] Henri Cartan, *Sur les rétractions d'une variété* (French, with English summary), C. R. Acad. Sci. Paris Sér. I Math. **303** (1986), no. 14, 715. MR870703

[CE] Henri Cartan and Samuel Eilenberg, *Homological algebra*, Princeton University Press, Princeton, NJ, 1956. MR77480

[ChE] J. Chaidez and O. Edtmair, *3D convex contact forms and the Ruelle invariant*, Invent. Math. **229** (2022), no. 1, 243–301, DOI 10.1007/s00222-022-01107-y. MR4438355

[ChH] Julian Chaidez and Michael Hutchings, *Computing Reeb dynamics on four-dimensional convex polytopes*, J. Comput. Dyn. **8** (2021), no. 4, 403–445, DOI 10.3934/jcd.2021016. MR4342232

[CDPT] Julian Chaidez, Ipsita Datta, Rohil Prasad, and Shira Tanny, *Contact homology and higher dimensional closing lemmas*, J. Mod. Dyn. **20** (2024), 67–153, DOI 10.3934/jmd.2024003. MR4729689

[CH] Daniel Cristofaro-Gardiner and Michael Hutchings, *From one Reeb orbit to two*, J. Differential Geom. **102** (2016), no. 1, 25–36. MR3447085

[CHP] Dan Cristofaro-Gardiner, Michael Hutchings, and Daniel Pomerleano, *Torsion contact forms in three dimensions have two or infinitely many Reeb orbits*, Geom. Topol. **23** (2019), no. 7, 3601–3645, DOI 10.2140/gt.2019.23.3601. MR4047649

[CHR] Daniel Cristofaro-Gardiner, Michael Hutchings, and Vinicius Gripp Barros Ramos, *The asymptotics of ECH capacities*, Invent. Math. **199** (2015), no. 1, 187–214, DOI 10.1007/s00222-014-0510-7. MR3294959

[Ch1] Marc Chaperon, *Une idée du type "géodésiques brisées" pour les systèmes hamiltoniens* (French, with English summary), C. R. Acad. Sci. Paris Sér. I Math. **298** (1984), no. 13, 293–296. MR765426

[Ch2] Marc Chaperon, *Quelques questions de géométrie symplectique (d'après, entre autres, Poincaré, Arnol′d, Conley et Zehnder)* (French), Bourbaki seminar, Vol. 1982/83, Astérisque, vol. 105, Soc. Math. France, Paris, 1983, pp. 231–249. MR728991

[CFH] K. Cieliebak, A. Floer, and H. Hofer, *Symplectic homology. II. A general construction*, Math. Z. **218** (1995), no. 1, 103–122, DOI 10.1007/BF02571891. MR1312580

[CFHW] K. Cieliebak, A. Floer, H. Hofer, and K. Wysocki, *Applications of symplectic homology. II. Stability of the action spectrum*, Math. Z. **223** (1996), no. 1, 27–45, DOI 10.1007/PL00004267. MR1408861

[CHLS] Kai Cieliebak, Helmut Hofer, Janko Latschev, and Felix Schlenk, *Quantitative symplectic geometry*, Dynamics, ergodic theory, and geometry, Math. Sci. Res. Inst. Publ., vol. 54, Cambridge Univ. Press, Cambridge, 2007, pp. 1–44, DOI 10.1017/CBO9780511755187.002. MR2369441

[CN] Katherine Christianson and Jo Nelson, *Symplectic embeddings of four-dimensional polydisks into balls*, Algebr. Geom. Topol. **18** (2018), no. 4, 2151–2178, DOI 10.2140/agt.2018.18.2151. MR3797064

[CGG] E. Cineli, V. Ginzburg and B. Gürel, Closed Orbits of Dynamically Convex Reeb Flows: Towards the HZ- and Multiplicity Conjectures, arxiv:2410.13093.

[Cl] Frank H. Clarke, *A classical variational principle for periodic Hamiltonian trajectories*, Proc. Amer. Math. Soc. **76** (1979), no. 1, 186–188, DOI 10.2307/2042942. MR534415

[C1] C. C. Conley, *Low energy transit orbits in the restricted three-body problem*, SIAM J. Appl. Math. **16** (1968), 732–746, DOI 10.1137/0116060. MR233535

[C2] Charles Conley, *Isolated invariant sets and the Morse index*, CBMS Regional Conference Series in Mathematics, vol. 38, American Mathematical Society, Providence, RI, 1978. MR511133

[CZ] C. C. Conley and E. Zehnder, *The Birkhoff-Lewis fixed point theorem and a conjecture of V. I. Arnol′d*, Invent. Math. **73** (1983), no. 1, 33–49, DOI 10.1007/BF01393824. MR707347

[CZ1] C. Conley and E. Zehnder, *Subharmonic solutions and Morse theory*, Phys. A **124** (1984), no. 1-3, 649–657, DOI 10.1016/0378-4371(84)90282-6. Mathematical physics, VII (Boulder, Colo., 1983). MR759212

[CZ2] Charles Conley and Eduard Zehnder, *Morse-type index theory for flows and periodic solutions for Hamiltonian equations*, Comm. Pure Appl. Math. **37** (1984), no. 2, 207–253, DOI 10.1002/cpa.3160370204. MR733717

[CZ3] C. Conley and E. Zehnder, *An index theory for periodic solutions of a Hamiltonian system*, Geometric dynamics (Rio de Janeiro, 1981), Lecture Notes in Math., vol. 1007, Springer, Berlin, 1983, pp. 132–145, DOI 10.1007/BFb0061415. MR730268

[Dol1] Albrecht Dold, *Lectures on algebraic topology*, 2nd ed., Grundlehren der Mathematischen Wissenschaften, vol. 200, Springer-Verlag, Berlin-New York, 1980. MR606196

[Dol2] Albrecht Dold, *The fixed point transfer of fibre-preserving maps*, Math. Z. **148** (1976), no. 3, 215–244, DOI 10.1007/BF01214520. MR433440

[Do] S. K. Donaldson, *An application of gauge theory to four-dimensional topology*, J. Differential Geom. **18** (1983), no. 2, 279–315. MR710056

[DK] S. K. Donaldson and P. B. Kronheimer, *The geometry of four-manifolds*, Oxford Mathematical Monographs, The Clarendon Press, Oxford University Press, New York, 1990. Oxford Science Publications. MR1079726

[Dou] Adrien Douady, *Nœuds et structures de contact en dimension* 3 *(d'après Daniel Bennequin)* (French), Bourbaki seminar, Vol. 1982/83, Astérisque, vol. 105, Soc. Math. France, Paris, 1983, pp. 129–148. MR728985

[Du] H. Scott Dumas, *The KAM story*, World Scientific Publishing Co. Pte. Ltd., Hackensack, NJ, 2014. A friendly introduction to the content, history, and significance of classical Kolmogorov-Arnold-Moser theory, DOI 10.1142/8955. MR3222196

[Ek1] Ivar Ekeland, *Une théorie de Morse pour les systèmes hamiltoniens convexes* (French, with English summary), Ann. Inst. H. Poincaré Anal. Non Linéaire **1** (1984), no. 1, 19–78. MR738494

[Ek2] Ivar Ekeland, *The best of all possible worlds*, University of Chicago Press, Chicago, IL, 2006. Mathematics and destiny; Translated from the 2000 French original. MR2259005

[EH1] I. Ekeland and H. Hofer, *Symplectic topology and Hamiltonian dynamics*, Math. Z. **200** (1989), no. 3, 355–378, DOI 10.1007/BF01215653. MR978597

[EH2] Ivar Ekeland and Helmut Hofer, *Symplectic topology and Hamiltonian dynamics. II*, Math. Z. **203** (1990), no. 4, 553–567, DOI 10.1007/BF02570756. MR1044064

[EH3] I. Ekeland and H. Hofer, *Convex Hamiltonian energy surfaces and their periodic trajectories*, Comm. Math. Phys. **113** (1987), no. 3, 419–469. MR925924

[EH4] I. Ekeland and H. Hofer, *Periodic solutions with prescribed minimal period for convex autonomous Hamiltonian systems*, Invent. Math. **81** (1985), no. 1, 155–188, DOI 10.1007/BF01388776. MR796195

[EL] Ivar Ekeland and Jean-Michel Lasry, *On the number of periodic trajectories for a Hamiltonian flow on a convex energy surface*, Ann. of Math. (2) **112** (1980), no. 2, 283–319, DOI 10.2307/1971148. MR592293

[ELa] I. Ekeland and L. Lassoued, *Multiplicité des trajectoires fermées de systèmes hamiltoniens convexes* (French, with English summary), Ann. Inst. H. Poincaré Anal. Non Linéaire **4** (1987), no. 4, 307–335. MR917740

[El] Halldór I. Elíasson, *Geometry of manifolds of maps*, J. Differential Geometry **1** (1967), 169–194. MR226681

[EnS] *Encyclopedia of the Solar System*, Edited by: Tilman Spohn, Doris Breuer and Torrence V. Johnson, ScienceDirect, Third Edition 2014.

[Eli] Y. Eliashberg, *Classification of overtwisted contact structures on 3-manifolds*, Invent. Math. **98** (1989), no. 3, 623–637, DOI 10.1007/BF01393840. MR1022310

[EGH] Y. Eliashberg, A. Givental, and H. Hofer, *Introduction to symplectic field theory*, Geom. Funct. Anal. **Special Volume** (2000), 560–673, DOI 10.1007/978-3-0346-0425-3_4. GAFA 2000 (Tel Aviv, 1999). MR1826267

[FR] Edward R. Fadell and Paul H. Rabinowitz, *Generalized cohomological index theories for Lie group actions with an application to bifurcation questions for Hamiltonian systems*, Invent. Math. **45** (1978), no. 2, 139–174, DOI 10.1007/BF01390270. MR478189

[Fe] Charles L. Fefferman, *The uncertainty principle*, Bull. Amer. Math. Soc. (N.S.) **9** (1983), no. 2, 129–206, DOI 10.1090/S0273-0979-1983-15154-6. MR707957

[FiH1] J. Fish and H. Hofer, Polyfold and SFT Notes I: A Primer on Polyfolds and Construction Tools, `arXiv:1806.07025`.

[FiH2] Joel W. Fish and Helmut Hofer, *Feral curves and minimal sets*, Ann. of Math. (2) **197** (2023), no. 2, 533–738, DOI 10.4007/annals.2023.197.2.2. MR4543440

[F0] Andreas Floer, *Morse theory for fixed points of symplectic diffeomorphisms*, Bull. Amer. Math. Soc. (N.S.) **16** (1987), no. 2, 279–281, DOI 10.1090/S0273-0979-1987-15517-0. MR876964

[F00] A. Floer, *Holomorphic curves and a Morse theory for fixed points of exact symplectomorphisms*, Aspects dynamiques et topologiques des groupes infinis de transformation de la mécanique (Lyon, 1986), Travaux en Cours, vol. 25, Hermann, Paris, 1987, pp. 49–60. MR906896

[F1] Andreas Floer, *The unregularized gradient flow of the symplectic action*, Comm. Pure Appl. Math. **41** (1988), no. 6, 775–813, DOI 10.1002/cpa.3160410603. MR948771

[F2] Andreas Floer, *Morse theory for Lagrangian intersections*, J. Differential Geom. **28** (1988), no. 3, 513–547. MR965228

[F3] Andreas Floer, *Witten's complex and infinite-dimensional Morse theory*, J. Differential Geom. **30** (1989), no. 1, 207–221. MR1001276

[F4] Andreas Floer, *A relative Morse index for the symplectic action*, Comm. Pure Appl. Math. **41** (1988), no. 4, 393–407, DOI 10.1002/cpa.3160410402. MR933228

[F5] Andreas Floer, *Proof of the Arnol′d conjecture for surfaces and generalizations to certain Kähler manifolds*, Duke Math. J. **53** (1986), no. 1, 1–32, DOI 10.1215/S0012-7094-86-05301-9. MR835793

[F6] Andreas Floer, *A refinement of the Conley index and an application to the stability of hyperbolic invariant sets*, Ergodic Theory Dynam. Systems **7** (1987), no. 1, 93–103, DOI 10.1017/S0143385700003825. MR886372

[F7] Andreas Floer, *Monopoles on asymptotically Euclidean 3-manifolds*, Bull. Amer. Math. Soc. (N.S.) **16** (1987), no. 1, 125–127, DOI 10.1090/S0273-0979-1987-15485-1. MR866030

[F8] A. Floer, *The configuration space of Yang-Mills-Higgs theory of asymptotically flat manifolds*, The Floer memorial volume, Progr. Math., vol. 133, Birkhäuser, Basel, 1995, pp. 43–75. MR1362822

[F9] A. Floer, *Monopoles on asymptotically flat manifolds*, The Floer memorial volume, Progr. Math., vol. 133, Birkhäuser, Basel, 1995, pp. 3–41. MR1362821

[F10] A. Floer, On the Arnold Conjecture, SUNY Otcober 29, 1986, video, `http://www.math.stonybrook.edu/Videos/Einstein/39-19861029-Floer.html`.

[F11] Andreas Floer, *Viterbo's index and the Morse index for the symplectic action*, Periodic solutions of Hamiltonian systems and related topics (Il Ciocco, 1986), NATO Adv. Sci. Inst. Ser. C: Math. Phys. Sci., vol. 209, Reidel, Dordrecht, 1987, pp. 147–152. MR920617

[F12] Andreas Floer, *Elliptic methods in variational problems*, Proceedings of the International Congress of Mathematicians, Vol. I, II (Kyoto, 1990), Math. Soc. Japan, Tokyo, 1991, pp. 87–97. MR1159207

[F13] Andreas Floer, *Symplectic fixed points and holomorphic spheres*, Comm. Math. Phys. **120** (1989), no. 4, 575–611. MR987770

[FH] A. Floer and H. Hofer, *Symplectic homology. I. Open sets in* $\mathbf{C}^n$, Math. Z. **215** (1994), no. 1, 37–88, DOI 10.1007/BF02571699. MR1254813

[FHS] Andreas Floer, Helmut Hofer, and Dietmar Salamon, *Transversality in elliptic Morse theory for the symplectic action*, Duke Math. J. **80** (1995), no. 1, 251–292, DOI 10.1215/S0012-7094-95-08010-7. MR1360618

[FHV] A. Floer, H. Hofer, and C. Viterbo, *The Weinstein conjecture in* $P \times \mathbf{C}^l$, Math. Z. **203** (1990), no. 3, 469–482, DOI 10.1007/BF02570750. MR1038712

[FHW] A. Floer, H. Hofer, and K. Wysocki, *Applications of symplectic homology. I*, Math. Z. **217** (1994), no. 4, 577–606, DOI 10.1007/BF02571962. MR1306027

[FW] Andreas Floer and Alan Weinstein, *Nonspreading wave packets for the cubic Schrödinger equation with a bounded potential*, J. Funct. Anal. **69** (1986), no. 3, 397–408, DOI 10.1016/0022-1236(86)90096-0. MR867665

[FZ] A. Floer and E. Zehnder, *Fixed point results for symplectic maps related to the Arnol′d conjecture*, Dynamical systems and bifurcations (Groningen, 1984), Lecture Notes in Math., vol. 1125, Springer, Berlin, 1985, pp. 47–63, DOI 10.1007/BFb0075634. MR798081

[Fo] Barry Fortune, *A symplectic fixed point theorem for* $\mathbf{C}\mathrm{P}^n$, Invent. Math. **81** (1985), no. 1, 29–46, DOI 10.1007/BF01388770. MR796189

[FoW] Barry Fortune and Alan Weinstein, *A symplectic fixed point theorem for complex projective spaces*, Bull. Amer. Math. Soc. (N.S.) **12** (1985), no. 1, 128–130, DOI 10.1090/S0273-0979-1985-15314-5. MR766969

[Fox] Ralph H. Fox, *On the Lusternik-Schnirelmann category*, Ann. of Math. (2) **42** (1941), 333–370, DOI 10.2307/1968905. MR4108

[Fra1] John Franks, *Generalizations of the Poincaré-Birkhoff theorem*, Ann. of Math. (2) **128** (1988), no. 1, 139–151, DOI 10.2307/1971464. MR951509

[Fra2] John Franks, *Area preserving homeomorphisms of open surfaces of genus zero*, New York J. Math. **2** (1996), 1–19, electronic. MR1371312

[FrLC] John Franks and Patrice Le Calvez, *Regions of instability for non-twist maps*, Ergodic Theory Dynam. Systems **23** (2003), no. 1, 111–141, DOI 10.1017/S0143385702000858. MR1971199

[Fran] Robert Franzosa, *Index filtrations and the homology index braid for partially ordered Morse decompositions*, Trans. Amer. Math. Soc. **298** (1986), no. 1, 193–213, DOI 10.2307/2000615. MR857439

[FKM] Urs Frauenfelder, Dayung Koh, and Agustin Moreno, *Symplectic methods in the numerical search of orbits in real-life planetary systems*, SIAM J. Appl. Dyn. Syst. **22** (2023), no. 4, 3284–3319, DOI 10.1137/22M1500459. MR4673680

[FvK] Urs Frauenfelder and Otto van Koert, *The restricted three-body problem and holomorphic curves*, Pathways in Mathematics, Birkhäuser/Springer, Cham, 2018, DOI 10.1007/978-3-319-72278-8. MR3837531

[FU] Daniel S. Freed and Karen K. Uhlenbeck, *Instantons and four-manifolds*, Mathematical Sciences Research Institute Publications, vol. 1, Springer-Verlag, New York, 1984, DOI 10.1007/978-1-4684-0258-2. MR757358

[Fr] Michael Hartley Freedman, *The topology of four-dimensional manifolds*, J. Differential Geometry **17** (1982), no. 3, 357–453. MR679066

[FuO] Kenji Fukaya and Kaoru Ono, *Arnold conjecture and Gromov-Witten invariant*, Topology **38** (1999), no. 5, 933–1048, DOI 10.1016/S0040-9383(98)00042-1. MR1688434

[FO^3_I] Kenji Fukaya, Yong-Geun Oh, Hiroshi Ohta, and Kaoru Ono, *Lagrangian intersection Floer theory: anomaly and obstruction. Part I*, AMS/IP Studies in Advanced Mathematics, vol. 46, American Mathematical Society, Providence, RI; International Press, Somerville, MA, 2009, DOI 10.1090/amsip/046.1. MR2553465

[FO^3_{II}] Kenji Fukaya, Yong-Geun Oh, Hiroshi Ohta, and Kaoru Ono, *Lagrangian intersection Floer theory: anomaly and obstruction. Part II*, AMS/IP Studies in Advanced Mathematics, vol. 46, American Mathematical Society, Providence, RI; International Press, Somerville, MA, 2009, DOI 10.1090/amsip/046.2. MR2548482

[GM] Sergei I. Gelfand and Yuri I. Manin, *Methods of homological algebra*, 2nd ed., Springer Monographs in Mathematics, Springer-Verlag, Berlin, 2003, DOI 10.1007/978-3-662-12492-5. MR1950475

[GG] Jean-Marc Ginoux and Christian Gerini, *Henri Poincaré*, World Scientific Publishing Co. Pte. Ltd., Hackensack, NJ, 2014. A biography through the daily papers; Translated from the French edition by Selma Naëck; With a foreword by Leon Chua. MR3308466

[Gi1] Viktor L. Ginzburg, *An embedding* $S^{2n-1} \to \mathbf{R}^{2n}$, $2n-1 \geq 7$, *whose Hamiltonian flow has no periodic trajectories*, Internat. Math. Res. Notices **2** (1995), 83–97, DOI 10.1155/S1073792895000079. MR1317645

[Gi2] Viktor L. Ginzburg, *The Conley conjecture*, Ann. of Math. (2) **172** (2010), no. 2, 1127–1180, DOI 10.4007/annals.2010.172.1129. MR2680488

[GiG1] Viktor L. Ginzburg and Başak Z. Gürel, *A* C^2*-smooth counterexample to the Hamiltonian Seifert conjecture in* $\mathbb{R}^4$, Ann. of Math. (2) **158** (2003), no. 3, 953–976, DOI 10.4007/annals.2003.158.953. MR2031857

[GiG2] Viktor L. Ginzburg and Başak Z. Gürel, *Conley conjecture revisited*, Int. Math. Res. Not. IMRN **3** (2019), 761–798, DOI 10.1093/imrn/rnx137. MR3910472

[Gom] Robert E. Gompf, *An infinite set of exotic* $\mathbf{R}^4$*'s*, J. Differential Geom. **21** (1985), no. 2, 283–300. MR816673

[GL] Joshua Evan Greene and Andrew Lobb, *The rectangular peg problem*, Ann. of Math. (2) **194** (2021), no. 2, 509–517, DOI 10.4007/annals.2021.194.2.4. MR4298749

[G] M. Gromov, *Pseudo holomorphic curves in symplectic manifolds*, Invent. Math. **82** (1985), no. 2, 307–347, DOI 10.1007/BF01388806. MR809718

[G1] Mikhael Gromov, *Partial differential relations*, Ergebnisse der Mathematik und ihrer Grenzgebiete (3) [Results in Mathematics and Related Areas (3)], vol. 9, Springer-Verlag, Berlin, 1986, DOI 10.1007/978-3-662-02267-2. MR864505

[G2] M. Gromov, Interview at the Simons foundation 18: Symplectic Geometry. `https://www.simonsfoundation.org/2014/12/22/mikhail-gromov/`.

[G3] M. Gromov, *Soft and hard symplectic geometry*, Proceedings of the International Congress of Mathematicians, Vol. 1, 2 (Berkeley, Calif., 1986), Amer. Math. Soc., Providence, RI, 1987, pp. 81–98. MR934217

[GuH] Jean Gutt, Jean Gutt, and Michael Hutchings, *Symplectic capacities from positive S^1-equivariant symplectic homology*, Algebr. Geom. Topol. **18** (2018), no. 6, 3537–3600, DOI 10.2140/agt.2018.18.3537. MR3868228

[HO] P. Haim-Kislev and Y. Ostrover, A Counterexample to Viterbo's Conjecture, `arXiv:2405.16513` (to appear Annals of Mathematics).

[Ha] Boris Hasselblatt, *Anatole Katok—a half-century of dynamics*, Notices Amer. Math. Soc. **66** (2019), no. 5, 708–719. MR3929470

[HaH] M. Hastings and J. Haah, Chain Complexes and Error-Correcting Codes, Volume 383, in 2021, 1621–1669.

[HeS] B. Helffer and J. Sjöstrand, *Puits multiples en mécanique semi-classique. IV. Étude du complexe de Witten* (French), Comm. Partial Differential Equations **10** (1985), no. 3, 245–340, DOI 10.1080/03605308508820379. MR780068

[He] Michel R. Herman, *Examples of compact hypersurfaces in $\mathbf{R}^{2p}$, $2p \geq 6$, with no periodic orbits*, Hamiltonian systems with three or more degrees of freedom (S'Agaró, 1995), NATO Adv. Sci. Inst. Ser. C: Math. Phys. Sci., vol. 533, Kluwer Acad. Publ., Dordrecht, 1999, pp. 126. MR1720888

[Hi] Nancy Hingston, *Subharmonic solutions of Hamiltonian equations on tori*, Ann. of Math. (2) **170** (2009), no. 2, 529–560, DOI 10.4007/annals.2009.170.529. MR2552101

[MH] Morris W. Hirsch, *Differential topology*, Graduate Texts in Mathematics, vol. 33, Springer-Verlag, New York, 1994. Corrected reprint of the 1976 original. MR1336822

[H0] H. Hofer, *Pseudoholomorphic curves in symplectizations with applications to the Weinstein conjecture in dimension three*, Invent. Math. **114** (1993), no. 3, 515–563, DOI 10.1007/BF01232679. MR1244912

[H1] Helmut Hofer, *Lagrangian embeddings and critical point theory*, Ann. Inst. H. Poincaré Anal. Non Linéaire **2** (1985), no. 6, 407–462. MR831040

[H2] H. Hofer, The Floer Jungle: 35 years of Floer Theory, `https://www.youtube.com/watch?v=kSNyU71MpgQ`.

[TE] `https://www.nytimes.com/2016/03/13/nyregion/death-in-texas-an-officers-odyssey.html`

[H4] Helmut Hofer, *On strongly indefinite functionals with applications*, Trans. Amer. Math. Soc. **275** (1983), no. 1, 185–214, DOI 10.2307/1999013. MR678344

[H5] Helmut Hofer, *A new proof for a result of Ekeland and Lasry concerning the number of periodic Hamiltonian trajectories on a prescribed energy surface* (English, with Italian summary), Boll. Un. Mat. Ital. B (6) **1** (1982), no. 3, 931–942. MR683483

[H6] Helmut Hofer, *The topological degree at a critical point of mountain-pass type*, Nonlinear functional analysis and its applications, Part 1 (Berkeley, Calif., 1983), Proc. Sympos. Pure Math., vol. 45, Amer. Math. Soc., Providence, RI, 1986, pp. 501–509, DOI 10.1090/pspum/045.1/843584. MR843584

[H7] H. Hofer, *Lusternik-Schnirelman-theory for Lagrangian intersections* (English, with French summary), Ann. Inst. H. Poincaré Anal. Non Linéaire **5** (1988), no. 5, 465–499. MR970850

[H8] H. Hofer, *On the topological properties of symplectic maps*, Proc. Roy. Soc. Edinburgh Sect. A **115** (1990), no. 1-2, 25–38, DOI 10.1017/S0308210500024549. MR1059642

[HS1] H. Hofer and D. A. Salamon, *Floer homology and Novikov rings*, The Floer memorial volume, Progr. Math., vol. 133, Birkhäuser, Basel, 1995, pp. 483–524. MR1362838

[FMem] The Floer memorial volume. Edited by Helmut Hofer, Clifford H. Taubes, Alan Weinstein and Eduard Zehnder. Progress in Mathematics, 133. Birkhäuser Verlag, Basel, 1995. xii+685 pp.

[HV] H. Hofer and C. Viterbo, *The Weinstein conjecture in the presence of holomorphic spheres*, Comm. Pure Appl. Math. **45** (1992), no. 5, 583–622, DOI 10.1002/cpa.3160450504. MR1162367

[HWZ] Helmut Hofer, Krzysztof Wysocki, and Eduard Zehnder, *Polyfold and Fredholm theory*, Ergebnisse der Mathematik und ihrer Grenzgebiete. 3. Folge. A Series of Modern Surveys in Mathematics [Results in Mathematics and Related Areas. 3rd Series. A Series of Modern Surveys in Mathematics], vol. 72, Springer, Cham, [2021] ©2021, DOI 10.1007/978-3-030-78007-4. MR4298268

[HWZ1] H. Hofer, K. Wysocki, and E. Zehnder, *The dynamics on three-dimensional strictly convex energy surfaces*, Ann. of Math. (2) **148** (1998), no. 1, 197–289, DOI 10.2307/120994. MR1652928

[HWZ2] H. Hofer, K. Wysocki, and E. Zehnder, *Finite energy foliations of tight three-spheres and Hamiltonian dynamics*, Ann. of Math. (2) **157** (2003), no. 1, 125–255, DOI 10.4007/annals.2003.157.125. MR1954266

[HZ0] H. Hofer and E. Zehnder, *Andreas Floer, 1956–1991*, Topological and geometrical methods in field theory (Turku, 1991), World Sci. Publ., River Edge, NJ, 1992, pp. xi. MR1224262

[HZ1] H. Hofer and E. Zehnder, *Periodic solutions on hypersurfaces and a result by C. Viterbo*, Invent. Math. **90** (1987), no. 1, 1–9, DOI 10.1007/BF01389030. MR906578

[HZ2] H. Hofer and E. Zehnder, Symplectic invariants and Hamiltonian dynamics. Birkhäuser Advanced Texts: Basler Lehrbücher. [Birkhäuser Advanced Texts: Basel Textbooks] Birkhäuser Verlag, Basel, 1994. xiv+341 pp.

[HZ3] H. Hofer and E. Zehnder, *A new capacity for symplectic manifolds*, Analysis, et cetera, Academic Press, Boston, MA, 1990, pp. 405–427. MR1039354

[Hu] Cole Hugelmeyer, *Inscribed rectangles in a smooth Jordan curve attain at least one third of all aspect ratios*, Ann. of Math. (2) **194** (2021), no. 2, 497–508, DOI 10.4007/annals.2021.194.2.3. MR4298748

[Hut1] Michael Hutchings, *An index inequality for embedded pseudoholomorphic curves in symplectizations*, J. Eur. Math. Soc. (JEMS) **4** (2002), no. 4, 313–361, DOI 10.1007/s100970100041. MR1941088

[Hut2] Michael Hutchings, *Quantitative embedded contact homology*, J. Differential Geom. **88** (2011), no. 2, 231–266. MR2838266

[Hut3] Michael Hutchings, *Embedded contact homology and its applications*, Proceedings of the International Congress of Mathematicians. Volume II, Hindustan Book Agency, New Delhi, 2010, pp. 1022–1041. MR2827830

[HutN1] Michael Hutchings and Jo Nelson, *Axiomatic S^1 Morse-Bott theory*, Algebr. Geom. Topol. **20** (2020), no. 4, 1641–1690, DOI 10.2140/agt.2020.20.1641. MR4127081

[HutN2] Michael Hutchings and Jo Nelson, *S^1-equivariant contact homology for hypertight contact forms*, J. Topol. **15** (2022), no. 3, 1455–1539, DOI 10.1112/topo.12240. MR4461852

[Ir1] Kei Irie, *Dense existence of periodic Reeb orbits and ECH spectral invariants*, J. Mod. Dyn. **9** (2015), 357–363, DOI 10.3934/jmd.2015.9.357. MR3436746

[Ir2] Kei Irie, *Equidistributed periodic orbits of C^∞-generic three-dimensional Reeb flows*, J. Symplectic Geom. **19** (2021), no. 3, 531–566, DOI 10.4310/JSG.2021.v19.n3.a2. MR4325412

[JQ] Arthur Jaffe and Frank Quinn, *"Theoretical mathematics": toward a cultural synthesis of mathematics and theoretical physics*, Bull. Amer. Math. Soc. (N.S.) **29** (1993), no. 1, 1–13, DOI 10.1090/S0273-0979-1993-00413-0. MR1202292

[JT] Arthur Jaffe and Clifford Taubes, *Vortices and monopoles*, Progress in Physics, vol. 2, Birkhäuser, Boston, MA, 1980. Structure of static gauge theories. MR614447

[Ku] Robert Franzosa, *Index filtrations and the homology index braid for partially ordered Morse decompositions*, Trans. Amer. Math. Soc. **298** (1986), no. 1, 193–213, DOI 10.2307/2000615. MR857439

[TTA] *T*ribute to Arnold, coordinating editors B. Khesin and S. Tabachnikov, `https://www.ams.org/notices/201203/rtx120300378p.pdf`.

[Ki] R. Kirby, Mathematical Review MR0710056 of [**Do**].

[Kl] Morris Kline, *Mathematical thought from ancient to modern times*, Oxford University Press, New York, 1972. MR472307

[La] Serge Lang, *Differential and Riemannian manifolds*, 3rd ed., Graduate Texts in Mathematics, vol. 160, Springer-Verlag, New York, 1995, DOI 10.1007/978-1-4612-4182-9. MR1335233

[LMc1] François Lalonde and Dusa McDuff, *The geometry of symplectic energy*, Ann. of Math. (2) **141** (1995), no. 2, 349–371, DOI 10.2307/2118524. MR1324138

[LMc2] François Lalonde and Dusa McDuff, *J-curves and the classification of rational and ruled symplectic 4-manifolds*, Contact and symplectic geometry (Cambridge, 1994), Publ. Newton Inst., vol. 8, Cambridge Univ. Press, Cambridge, 1996, pp. 3–42. MR1432456

[LMc3] François Lalonde and Dusa McDuff, *The classification of ruled symplectic 4-manifolds*, Math. Res. Lett. **3** (1996), no. 6, 769–778, DOI 10.4310/MRL.1996.v3.n6.a5. MR1426534

[LMc4] François Lalonde and Dusa McDuff, *Hofer's L^∞-geometry: energy and stability of Hamiltonian flows. I, II*, Invent. Math. **122** (1995), no. 1, 1–33, 35–69, DOI 10.1007/BF01231437. MR1354953

[LR] D. G. Larman and C. A. Rogers, *The existence of a centrally symmetric convex body with central sections that are unexpectedly small*, Mathematika **22** (1975), no. 2, 164–175, DOI 10.1112/S0025579300006033. MR390914

[LS] François Laudenbach and Jean-Claude Sikorav, *Persistance d'intersection avec la section nulle au cours d'une isotopie hamiltonienne dans un fibré cotangent* (French), Invent. Math. **82** (1985), no. 2, 349–357, DOI 10.1007/BF01388807. MR809719

[Lee] John M. Lee, *Introduction to smooth manifolds*, 2nd ed., Graduate Texts in Mathematics, vol. 218, Springer, New York, 2013. MR2954043

[LT] Gang Liu and Gang Tian, *Weinstein conjecture and GW-invariants*, Commun. Contemp. Math. **2** (2000), no. 4, 405–459, DOI 10.1142/S0219199700000256. MR1806943

[LoMcO] Robert B. Lockhart and Robert C. McOwen, *Elliptic differential operators on noncompact manifolds*, Ann. Scuola Norm. Sup. Pisa Cl. Sci. (4) **12** (1985), no. 3, 409–447. MR837256

[LZ] Yiming Long and Chaofeng Zhu, *Closed characteristics on compact convex hypersurfaces in* $\mathbf{R}^{2n}$, Ann. of Math. (2) **155** (2002), no. 2, 317–368, DOI 10.2307/3062120. MR1906590

[Ly] A. Lyapunov, Sur les figures d'équilibre peu différents des ellipsoides d'une masse liquide homogène douée d'un mouvement de rotation, Zap. Akad. Nauk St. Petersburg (1906), 1–225.

[MaT] Ib Madsen and Jørgen Tornehave, *From calculus to cohomology*, Cambridge University Press, Cambridge, 1997. de Rham cohomology and characteristic classes. MR1454127

[Mas] William S. Massey, *Homology and cohomology theory*, Monographs and Textbooks in Pure and Applied Mathematics, Vol. 46, Marcel Dekker, Inc., New York-Basel, 1978. An approach based on Alexander-Spanier cochains. MR488016

[Mau] Charles Richard Francis Maunder, *Algebraic topology*, Cambridge University Press, Cambridge-New York, 1980. MR694843

[McD0] `https://www.simonsfoundation.org/2014/06/23/dusa-mcduff/`

[McD1] Dusa McDuff, *Examples of symplectic structures*, Invent. Math. **89** (1987), no. 1, 13–36, DOI 10.1007/BF01404672. MR892186

[McD2] Dusa McDuff, *The structure of rational and ruled symplectic 4-manifolds*, J. Amer. Math. Soc. **3** (1990), no. 3, 679–712, DOI 10.2307/1990934. MR1049697

[McD3] Dusa McDuff, *Elliptic methods in symplectic geometry*, Bull. Amer. Math. Soc. (N.S.) **23** (1990), no. 2, 311–358, DOI 10.1090/S0273-0979-1990-15928-2. MR1039425

[McD4] Dusa McDuff, *The Hofer conjecture on embedding symplectic ellipsoids*, J. Differential Geom. **88** (2011), no. 3, 519–532. MR2844441

[McDS] Dusa McDuff and Dietmar Salamon, *J-holomorphic curves and symplectic topology*, American Mathematical Society Colloquium Publications, vol. 52, American Mathematical Society, Providence, RI, 2004, DOI 10.1090/coll/052. MR2045629

[McWQ] A Fight to Fix Geometry's Foundations, `https://www.quantamagazine.org/the-fight-to-fix-symplectic-geometry-20170209/`

[Mi] John Milnor, *On manifolds homeomorphic to the 7-sphere*, Ann. of Math. (2) **64** (1956), 399–405, DOI 10.2307/1969983. MR82103

[Mi2] John Milnor, *Lectures on the h-cobordism theorem*, Princeton University Press, Princeton, NJ, 1965. Notes by L. Siebenmann and J. Sondow. MR190942

[Mi3] J. Milnor, *Morse theory*, Annals of Mathematics Studies, No. 51, Princeton University Press, Princeton, NJ, 1963. Based on lecture notes by M. Spivak and R. Wells. MR163331

[Mo1] Marston Morse, *Relations between the critical points of a real function of n independent variables*, Trans. Amer. Math. Soc. **27** (1925), no. 3, 345–396, DOI 10.2307/1989110. MR1501318

[Mo2] Marston Morse, *The calculus of variations in the large*, American Mathematical Society Colloquium Publications, vol. 18, American Mathematical Society, Providence, RI, 1996. Reprint of the 1932 original, DOI 10.1090/coll/018. MR1451874

[Mo3] Marston Morse and Stewart S. Cairns, *Critical point theory in global analysis and differential topology: An introduction*, Pure and Applied Mathematics, Vol. 33, Academic Press, New York-London, 1969. MR245046

[Mo4] Marston Morse, *George David Birkhoff and his mathematical work*, Bull. Amer. Math. Soc. **52** (1946), 357–391, DOI 10.1090/S0002-9904-1946-08553-5. MR16341

[Mos0] Jürgen Moser, *Stable and random motions in dynamical systems*, Annals of Mathematics Studies, No. 77, Princeton University Press, Princeton, NJ; University of Tokyo Press, Tokyo, 1973. With special emphasis on celestial mechanics; Hermann Weyl Lectures, the Institute for Advanced Study, Princeton, N. J. MR442980

[Mos] J. Moser, *Periodic orbits near an equilibrium and a theorem by Alan Weinstein*, Comm. Pure Appl. Math. **29** (1976), no. 6, 724–747, DOI 10.1002/cpa.3160290613. MR426052

[JvN] Johann von Neumann, *Mathematische Grundlagen der Quantenmechanik* (German), Die Grundlehren der mathematischen Wissenschaften, Band 38, Springer-Verlag, Berlin-New York, 1968. Unveränderter Nachdruck der ersten Auflage von 1932. MR223138

[Noe1] Emmy Noether, *Gesammelte Abhandlungen/Collected papers* (German), Springer Collected Works in Mathematics, Springer, Heidelberg, 2013. Edited and with an introduction by Nathan Jacobson; With an introductory address by P. S. Alexandrov; Reprint of the 1983 edition. MR3114982

[Noe2] Emmy Noether, *Invariant variation problems*, Transport Theory Statist. Phys. **1** (1971), no. 3, 186–207, DOI 10.1080/00411457108231446. MR406752

[NFA] Felix E. Browder (ed.), *Nonlinear functional analysis and its applications. Part 2*, Proceedings of Symposia in Pure Mathematics, vol. 45, American Mathematical Society, Providence, RI, 1986, DOI 10.1090/pspum/045.2. MR843589

[Not] *Karen Uhlenbeck awarded Abel prize*, Notices Amer. Math. Soc. **66** (2019), no. 6, 939–941. MR3929583

[Lui] S. H. Lui, *An interview with Vladimir Arnol′d*, Notices Amer. Math. Soc. **44** (1997), no. 4, 432–438. MR1435453

[Pa] John Pardon, *An algebraic approach to virtual fundamental cycles on moduli spaces of pseudo-holomorphic curves*, Geom. Topol. **20** (2016), no. 2, 779–1034, DOI 10.2140/gt.2016.20.779. MR3493097

[PS] R. S. Palais and S. Smale, *A generalized Morse theory*, Bull. Amer. Math. Soc. **70** (1964), 165–172, DOI 10.1090/S0002-9904-1964-11062-4. MR158411

[Per] T. Perutz, Math Review MR2459307 (2010k:53153).

[Pi] Albrecht Pietsch, *History of Banach spaces and linear operators*, Birkhäuser Boston, Inc., Boston, MA, 2007. MR2300779

[Pol] Leonid Polterovich, *The geometry of the group of symplectic diffeomorphisms*, Lectures in Mathematics ETH Zürich, Birkhäuser Verlag, Basel, 2001, DOI 10.1007/978-3-0348-8299-6. MR1826128

[Pro] Felix E. Browder (ed.), *Nonlinear functional analysis and its applications. Part 2*, Proceedings of Symposia in Pure Mathematics, vol. 45, American Mathematical Society, Providence, RI, 1986, DOI 10.1090/pspum/045.2. MR843589

[QC] Quantized Columns, Topology 101: The Hole Truth, Quanta Magazine, `https://www.quantamagazine.org/topology-101-how-mathematicians-study-holes-20210126/`.

[Ra0] P. H. Rabinowitz, *Periodic solutions of nonlinear hyperbolic partial differential equations*, Comm. Pure Appl. Math. **20** (1967), 145–205, DOI 10.1002/cpa.3160200105. MR206507

[Ra1] Paul H. Rabinowitz, *Free vibrations for a semilinear wave equation*, Comm. Pure Appl. Math. **31** (1978), no. 1, 31–68, DOI 10.1002/cpa.3160310103. MR470378

[Ra2] Paul H. Rabinowitz, *Periodic solutions of Hamiltonian systems*, Comm. Pure Appl. Math. **31** (1978), no. 2, 157–184, DOI 10.1002/cpa.3160310203. MR467823

[Ra3] Paul H. Rabinowitz, *Critical points of indefinite functionals and periodic solutions of differential equations*, Proceedings of the International Congress of Mathematicians (Helsinki, 1978), Acad. Sci. Fennica, Helsinki, 1980, pp. 791–796. MR562689

[Ra4] Paul H. Rabinowitz, *Periodic solutions of a Hamiltonian system on a prescribed energy surface*, J. Differential Equations **33** (1979), no. 3, 336–352, DOI 10.1016/0022-0396(79)90069-X. MR543703

[Ra5] Paul H. Rabinowitz, *Minimax methods for indefinite functionals*, Nonlinear functional analysis and its applications, Part 2 (Berkeley, Calif., 1983), Proc. Sympos. Pure Math., vol. 45, Amer. Math. Soc., Providence, RI, 1986, pp. 287–306, DOI 10.1090/pspum/045.2/843616. MR843616

[RAEZ] P. H. Rabinowitz, A. Ambrosetti, I. Ekeland, and E. J. Zehnder (eds.), *Periodic solutions of Hamiltonian systems and related topics*, NATO Advanced Science Institutes Series C: Mathematical and Physical Sciences, vol. 209, D. Reidel Publishing Co., Dordrecht, 1987, DOI 10.1007/978-94-009-3933-2. MR920604

[Re] S. Rezchikov, Integral Arnol'd Conjecture, arxiv2209.11165, `arXiv:2209.11165`.

[RS] Joel Robbin and Dietmar Salamon, *The Maslov index for paths*, Topology **32** (1993), no. 4, 827–844, DOI 10.1016/0040-9383(93)90052-W. MR1241874

[RS1] Joel Robbin and Dietmar Salamon, *The spectral flow and the Maslov index*, Bull. London Math. Soc. **27** (1995), no. 1, 1–33, DOI 10.1112/blms/27.1.1. MR1331677

[RS2] Joel W. Robbin and Dietmar Salamon, *Dynamical systems, shape theory and the Conley index*, Ergodic Theory Dynam. Systems **8*** (1988), no. Charles Conley Memorial Issue, 375–393, DOI 10.1017/S0143385700009494. MR967645

[Ry] Krzysztof P. Rybakowski, *The homotopy index and partial differential equations*, Universitext, Springer-Verlag, Berlin, 1987, DOI 10.1007/978-3-642-72833-4. MR910097

[Ru] Walter Rudin, *Functional analysis*, 2nd ed., International Series in Pure and Applied Mathematics, McGraw-Hill, Inc., New York, 1991. MR1157815

[SR] Siobhan Roberts, *Genius at play*, Bloomsbury Press, New York, 2015. The curious mind of John Horton Conway. MR3329687

[SaU] J. Sacks and K. Uhlenbeck, *The existence of minimal immersions of 2-spheres*, Ann. of Math. (2) **113** (1981), no. 1, 1–24, DOI 10.2307/1971131. MR604040

[Sa1] Dietmar Salamon, *Morse theory, the Conley index and Floer homology*, Bull. London Math. Soc. **22** (1990), no. 2, 113–140, DOI 10.1112/blms/22.2.113. MR1045282

[Sa2] Dietmar Salamon, *Connected simple systems and the Conley index of isolated invariant sets*, Trans. Amer. Math. Soc. **291** (1985), no. 1, 1–41, DOI 10.2307/1999893. MR797044

[Sa3] Dietmar Salamon, *Lectures on Floer homology*, Symplectic geometry and topology (Park City, UT, 1997), IAS/Park City Math. Ser., vol. 7, Amer. Math. Soc., Providence, RI, 1999, pp. 143–229, DOI 10.1016/S0165-2427(99)00127-0. MR1702944

[Sa] Arthur Sard, *The measure of the critical values of differentiable maps*, Bull. Amer. Math. Soc. **48** (1942), 883–890, DOI 10.1090/S0002-9904-1942-07811-6. MR7523

[Schm] Erhard Schmidt, *Zur Theorie der linearen und nichtlinearen Integralgleichungen. III. Teil* (German), Math. Ann. **65** (1908), no. 3, 370–399, DOI 10.1007/BF01456418. MR1511472

[Sch] Matthias Schwarz, *Morse homology*, Progress in Mathematics, vol. 111, Birkhäuser Verlag, Basel, 1993, DOI 10.1007/978-3-0348-8577-5. MR1239174

[Sei] Paul Seidel, *A biased view of symplectic cohomology*, Current developments in mathematics, 2006, Int. Press, Somerville, MA, 2008, pp. 211–253. MR2459307

[Se] H. Seifert, *Periodische Bewegungen mechanischer Systeme* (German), Math. Z. **51** (1948), 197–216, DOI 10.1007/BF01291002. MR25693

[Si] Jean-Claude Sikorav, *Points fixes d'une application symplectique homologue à l'identité* (French), J. Differential Geom. **22** (1985), no. 1, 49–79. MR826424

[Sim] Carl P. Simon, *A bound for the fixed-point index of an area-preserving map with applications to mechanics*, Invent. Math. **26** (1974), 187–200, DOI 10.1007/BF01418948. MR353372

[Sm1] S. Smale, *Differentiable dynamical systems*, Bull. Amer. Math. Soc. **73** (1967), 747–817, DOI 10.1090/S0002-9904-1967-11798-1. MR228014

[Sm2] S. Smale, *An infinite dimensional version of Sard's theorem*, Amer. J. Math. **87** (1965), 861–866, DOI 10.2307/2373250. MR185604

[Sp] Edwin H. Spanier, *Cohomology theory for general spaces*, Ann. of Math. (2) **49** (1948), 407–427, DOI 10.2307/1969289. MR24621

[St] Michael Struwe, *Variational methods*, 4th ed., Ergebnisse der Mathematik und ihrer Grenzgebiete. 3. Folge. A Series of Modern Surveys in Mathematics [Results in Mathematics and Related Areas. 3rd Series. A Series of Modern Surveys in Mathematics], vol. 34, Springer-Verlag, Berlin, 2008. Applications to nonlinear partial differential equations and Hamiltonian systems. MR2431434

[St1] Michael Struwe, *Existence of periodic solutions of Hamiltonian systems on almost every energy surface*, Bol. Soc. Brasil. Mat. (N.S.) **20** (1990), no. 2, 49–58, DOI 10.1007/BF02585433. MR1143173

[Su] M. G. Sullivan, *K-theoretic invariants for Floer homology*, Geom. Funct. Anal. **12** (2002), no. 4, 810–872, DOI 10.1007/s00039-002-8267-3. MR1935550

[T-Ar] *Tribute to Vladimir Arnold*, Notices Amer. Math. Soc. **59** (2012), no. 3, 378–399, DOI 10.1090/noti810. Boris Khesin and Serge Tabachnikov, coordinating editors. MR2931629

[M-Ar] *Memories of Vladimir Arnold*, Notices Amer. Math. Soc. **59** (2012), no. 4, 482–502, DOI 10.1090/noti813. Boris Khesin and Serge Tabachnikov, coordinating editors. MR2951953

[Tau] Clifford Henry Taubes, *Self-dual Yang-Mills connections on non-self-dual 4-manifolds*, J. Differential Geometry **17** (1982), no. 1, 139–170. MR658473

[Tau1] Clifford Henry Taubes, *The Seiberg-Witten equations and the Weinstein conjecture*, Geom. Topol. **11** (2007), 2117–2202, DOI 10.2140/gt.2007.11.2117. MR2350473

[Tau2] Clifford Henry Taubes, *Stability in Yang-Mills theories*, Comm. Math. Phys. **91** (1983), no. 2, 235–263. MR723549

[Th] René Thom, *Sur une partition en cellules associée à une fonction sur une variété* (French), C. R. Acad. Sci. Paris **228** (1949), 973–975. MR29160

[Tr] Hans Triebel, *Interpolation theory, function spaces, differential operators*, North-Holland Mathematical Library, vol. 18, North-Holland Publishing Co., Amsterdam-New York, 1978. MR503903

[Tu] Loring W. Tu, *An introduction to manifolds*, 2nd ed., Universitext, Springer, New York, 2011, DOI 10.1007/978-1-4419-7400-6. MR2723362

[Uh1] Karen K. Uhlenbeck, *Removable singularities in Yang-Mills fields*, Comm. Math. Phys. **83** (1982), no. 1, 11–29. MR648355

[Uh2] Karen K. Uhlenbeck, *Connections with L^p bounds on curvature*, Comm. Math. Phys. **83** (1982), no. 1, 31–42. MR648356

[Vai] Edward Witten, *Supersymmetry and Morse theory*, J. Differential Geometry **17** (1982), no. 4, 661–692 (1983). MR683171

[Ve] I. N. Vekua, *Generalized analytic functions*, Pergamon Press, London-Paris-Frankfurt; Addison-Wesley Publishing Company, Inc., Reading, MA, 1962. MR150320

[Vic] James W. Vick, *Homology theory*, Pure and Applied Mathematics, Vol. 53, Academic Press, New York-London, 1973. An introduction to algebraic topology. MR375279

[Vi1] Claude Viterbo, *A proof of Weinstein's conjecture in* $\mathbf{R}^{2n}$ (English, with French summary), Ann. Inst. H. Poincaré Anal. Non Linéaire **4** (1987), no. 4, 337–356. MR917741

[Vi2] Claude Viterbo, *Symplectic topology as the geometry of generating functions*, Math. Ann. **292** (1992), no. 4, 685–710, DOI 10.1007/BF01444643. MR1157321

[Vi3] Claude Viterbo, *Metric and isoperimetric problems in symplectic geometry*, J. Amer. Math. Soc. **13** (2000), no. 2, 411–431, DOI 10.1090/S0894-0347-00-00328-3. MR1750956

[Vi4] C. Viterbo, *Functors and computations in Floer homology with applications. I*, Geom. Funct. Anal. **9** (1999), no. 5, 985–1033, DOI 10.1007/s000390050106. MR1726235

[Vol] Voltaire, Candide, (translated by Lowell Bair), Bantam Books, New York, Toronto, London, Sydney and Auckland.

[Wa] Frank W. Warner, *Foundations of differentiable manifolds and Lie groups*, Graduate Texts in Mathematics, vol. 94, Springer-Verlag, New York-Berlin, 1983. Corrected reprint of the 1971 edition. MR722297

[Weib] Charles A. Weibel, *History of homological algebra*, History of topology, North-Holland, Amsterdam, 1999, pp. 797–836, DOI 10.1016/B978-044482375-5/50029-8. MR1721123

[We1] Alan Weinstein, *Perturbation of periodic manifolds of Hamiltonian systems*, Bull. Amer. Math. Soc. **77** (1971), 814–818, DOI 10.1090/S0002-9904-1971-12816-1. MR281230

[We2] Alan Weinstein, *Symplectic manifolds and their Lagrangian submanifolds*, Advances in Math. **6** (1971), 329–346 (1971), DOI 10.1016/0001-8708(71)90020-X. MR286137

[We3] Alan Weinstein, *Lagrangian submanifolds and Hamiltonian systems*, Ann. of Math. (2) **98** (1973), 377–410, DOI 10.2307/1970911. MR331428

[We4] Alan Weinstein, *Periodic orbits for convex Hamiltonian systems*, Ann. of Math. (2) **108** (1978), no. 3, 507–518, DOI 10.2307/1971185. MR512430

[We5] Alan Weinstein, *Symplectic geometry*, Bull. Amer. Math. Soc. (N.S.) **5** (1981), no. 1, 1–13, DOI 10.1090/S0273-0979-1981-14911-9. MR614310

[We6] Alan Weinstein, *On the hypotheses of Rabinowitz' periodic orbit theorems*, J. Differential Equations **33** (1979), no. 3, 353–358, DOI 10.1016/0022-0396(79)90070-6. MR543704

[We7] Alan Weinstein, *Nonlinear stabilization of quasimodes*, Geometry of the Laplace operator (Proc. Sympos. Pure Math., Univ. Hawaii, Honolulu, Hawaii, 1979), Proc. Sympos. Pure Math., XXXVI, Amer. Math. Soc., Providence, RI, 1980, pp. 301–318. MR573443

[Wey] Hermann Weyl, *The Classical Groups. Their Invariants and Representations*, Princeton University Press, Princeton, NJ, 1939. MR255

[Wi] Edward Witten, *Supersymmetry and Morse theory*, J. Differential Geometry **17** (1982), no. 4, 661–692 (1983). MR683171

[Ze0] Eduard Zehnder, *The beginnings of symplectic topology in Bochum in the early eighties*, Jahresber. Dtsch. Math.-Ver. **121** (2019), no. 2, 71–90, DOI 10.1365/s13291-019-00201-0. MR3955759

[Zei] Eberhard Zeidler, *Nonlinear functional analysis and its applications. I*, Springer-Verlag, New York, 1986. Fixed-point theorems; Translated from the German by Peter R. Wadsack, DOI 10.1007/978-1-4612-4838-5. MR816732

Index of Names

General Index

Selected Published Titles in This Series

For a complete list of titles in this series, visit the
AMS Bookstore at **www.ams.org/bookstore/hmathseries/**.

FSC
www.fsc.org
MIX
Papier | Fördert
gute Waldnutzung
FSC® C083411

Kolibri 360 GmbH
Ferdinand-Jühlke-Straße 7
99095 Erfurt, Deutschland
produktsicherheit@kolibri360.de